The Correspondence of Charles Hutton (1737–1823)

The Correspondence of Charles Hutton (1737–1823)

Mathematical Networks in Georgian Britain

Editor

BENJAMIN WARDHAUGH

OXFORD
UNIVERSITY PRESS

OXFORD
UNIVERSITY PRESS

Great Clarendon Street, Oxford, OX2 6DP,
United Kingdom

Oxford University Press is a department of the University of Oxford. It furthers the University's objective of excellence in research, scholarship, and education by publishing worldwide. Oxford is a registered trade mark of Oxford University Press in the UK and in certain other countries

First Edition published in 2017

Published in the United States of America by Oxford University Press
198 Madison Avenue, New York, NY 10016, United States of America

British Library Cataloguing in Publication Data
Data available

Library of Congress Control Number: 2017943714

ISBN 978–0–19–880504–5

ACKNOWLEDGEMENTS

I am grateful to the following for permission to publish the letters in this volume: the Bodleian Libraries, University of Oxford; the British Library Board; Durham County Record Office; the Devonshire Collections, Chatsworth; East Sussex Record Office; Edinburgh City Archives; Edinburgh University Main Library; Harry Ransom Center, the University of Texas at Austin; Harvard University; the Library of Birmingham; the Master and Fellows of Trinity College Cambridge; the Mathematical Association; the National Archive, Kew; the National Library of Scotland; Newcastle Literary and Philosophical Society; Oxford Museum of the History of Science; Plymouth and West Devon Record Office; Pollinger Limited (www.pollingerltd.com) on behalf of Lovelace Byron Papers; Portsmouth Library and Archive Service, Portsmouth City Council; the Royal Astronomical Society; the Royal Society; Senate House Library, London; the Smithsonian Institution; the Syndics of Cambridge University Library; University College, London; and the Wellcome Collection.

I am grateful to the librarians and other staff at those institutions who assisted with access to rare books and manuscripts and in a number of cases supplied photographs of items in their collections. Particular thanks are due to Jessica Wardhaugh for her advice about the French texts of Letters 16, 24, 66, 93, and 120.

The research for this volume was funded by a grant from the Arts and Humanities Research Council, and supported by All Souls College and the History Faculty at the University of Oxford.

Finally, I am immensely grateful to Emily Winterburn and Shelley Costa for undertaking to read the text of this volume before publication. Errors and infelicities which remain are my own responsibility alone.

CONTENTS

INTRODUCTION

Charles Hutton (1737–1823) was one of the most prominent British mathematicians of his generation, possibly the most prominent. His name is attached to no major discovery or innovation, and indeed he was known in his lifetime more for his expository and popular writing, and for his teaching, than for his production of new mathematics or science, notwithstanding the fact that in 1778 he received the Royal Society's Copley Medal for his work on the force of gunpowder. He is of interest instead because of the variety of roles he played in British mathematics, the dexterity with which he navigated, exploited, and shaped personal and intellectual networks in mathematics and natural philosophy, and the length and visibility of his career. Equally, his correspondence is full of interest because it bears direct witness to the overlapping, the intersection and interaction of the different networks in which Hutton moved. Hutton and his letters allow us to see something of how mathematical networks and relationships worked, in ways which can be very difficult using printed sources alone. Since his surviving letters are widely scattered, we can only do this by bringing them together in a single collected volume.

Hutton's biography is discussed at length and fully documented in Wardhaugh, *Gunpowder and Geometry*. The present introduction sets out Hutton's life in outline, paying particular attention to the different networks in which he participated. Hutton moved in a number of different worlds, and it is a purpose of publishing this admittedly fragmentary correspondence that it illustrates the degree to which those worlds overlapped and the ways an individual could exploit them to promote himself and his work. Hutton was a mathematical practitioner and a philomath. He was a Fellow of the Royal Society and a mathematical popularizer. He was part of the Greenwich-based network around Nevil Maskelyne and of the nonconformist network that linked Benjamin Franklin and others with the founders of the Newcastle Literary and Philosophical Society. These various links cannot be cleanly disentangled—certainly not using the materials we have—but a broadly chronological approach helps to see some of the patterns in an eclectic and piecemeal career, and it helps us to understand how Hutton, remarkably, was able to thrive and continue to thrive after being driven out of the most prestigious scientific institution of his day. My intention in this introduction is to build on recent work such as that of Emily Winterburn in displaying something of the complexity and richness of the networks of support and publicity that eighteenth-century natural philosophers and mathematicians drew on and with which they worked.

Figure 1: Hutton, *Tracts* (1812): frontispiece. Oxford, Bodleian Library, 8 Y 98–100 Art.BS, reproduced by permission of the Bodleian Library.

Biography

Newcastle and Longbenton

Charles Hutton was born in Newcastle-upon-Tyne in 1737 to parents who may have been immigrants to the area. He was the youngest of several brothers and had at least two sisters. His father occupied a position of trust in the estates of Lord Ravenscroft; some sources say land steward, others viewer or overman of mines. He died in 1742 and his widow remarried; she in turn died in 1760. Her new husband was an overman at the Rose Pit at Longbenton colliery on north Tyneside; Hutton worked as a hewer there for a few months in 1755–6.[1]

Hutton's family stood fairly low in the spectrum of access to education in this period, but by no means at its very bottom; although his brothers seem to have received only the most minimal schooling, Hutton himself was kept at school rather longer, in Newcastle and in and around Longbenton. A sense of his talents, as well as possibly a disability resulting from a childhood accident, led to his being offered the chance to leave the mines in his late teens and take charge of a local school when its teacher moved away. Probably by late 1756 he was teaching there, and soon afterwards at a larger schoolroom in nearby Stote Hall. A series of moves followed through at least three schools in Newcastle itself, the last a purpose-built establishment on the fashionable Westgate Street.[2] Hutton throve as a teacher and was able to command high prices; his school became a venue for teacher training and for popular lectures.[3] He published an arithmetic textbook and an encyclopaedic treatment of geometry, surveyed the city of Newcastle, and edited a five-volume collection of mathematical and other material from the popular periodical *The Ladies' Diary*.[4]

These relatively conventional markers of individual success also illustrate Hutton's membership of a network of mathematical teachers and practitioners based on Newcastle and the north-east of England. When local inventor John Richmond had a scheme to promote connected with gauging, he wrote to scientific lecturer John Rotheram, schoolmaster Charles Hutton, natural philosopher and teacher Robert Harrison, and mathematician William Emerson, in that order: all were based in or around Newcastle.[5] Other members of the network included Hutton's assistants George Coughron and John Fryer, and there were plenty of others in the region who enjoyed more or less strong ties to one another: schoolmasters, surveyors, authors, and printers of textbooks.[6] Thomas Saint, for instance, the publisher of the *Newcastle Courant* in which Hutton's early school advertisements appeared, was also the publisher of Hutton's first book in its second edition, the author of the only arithmetic textbook published in Newcastle before Hutton's and the editor of a mathematics column in the *Courant.*

These were mathematical practitioners in E.G.R. Taylor's sense: men and the occasional woman who made their living from mathematical teaching, writing, and printing, as well as surveying, gauging, and other mathematical activities.[7] To some degree they were united as mutual competitors when they were not involved in direct relationships as—for instance—assistants and publishers. Broadly similar communities of mathematical practitioners had existed certainly in the seventeenth century and perhaps as early as the sixteenth; changes in the range and complexity of mathematical instruments in anything like common use, changes in the frequency with which estate surveys were carried out, and changes in the demand for and nature of mathematical schools had all made their mark over that long period, but much was not substantially altered.

It was a network which, importantly, had a function in the transmission of skills; Winterburn has pointed out that those who were and are sometimes called self-educated (and who sometimes described themselves in that way) often learned their skills from peers and older players within networks of this kind.[8] Hutton learned mathematics first, as far as we know, from a local school and then from his teacher Jonathan Ivison; he received the crucial break that made him a teacher rather than a miner from Ivison. While teaching, he attended night classes in Newcastle with a Mr James; and while writing his treatise on mensuration he regularly visited a Mr Young at Prudhoe: riding across for the weekend, it was later said, to work over drafts of the book together.[9] Thus, when Hutton's biographers and obituarists represented him as educating himself through a remorseless programme of reading during his Newcastle years they did not tell the whole story; much that he gained came directly from people, not through their books.[10]

It was also, in part, personal ties and personal endorsements that enabled Hutton to flourish as a teacher. As well as the good will of Ivison, Hutton enjoyed that of another locally prominent family, who allowed him the use of a room in Stote Hall in which to teach when his classes outgrew their first premises at Jesmond. Another supporter was local landowner Robert Shaftoe, who engaged Hutton to teach mathematics to his children, and subsequently gave him access to his personal library and paid for the printing of his first book. Once established in Newcastle, Hutton developed a good relationship with the local grammar school, and students from the school were sent to him for mathematical teaching.[11] All these items of personal goodwill—of patronage—helped his rise as a teacher.

As well as the local mathematical practitioners' network and its patrons, it is conspicuous that Hutton was at the same time a participant in the rather different but overlapping national network of the philomaths. This was a more recent phenomenon; it had been conjured into existence early in the eighteenth century by the *Ladies' Diary* and continued well into the nineteenth in that forum and in a number of other national and local periodicals which printed problems and solutions from their readers and—usually—offered prizes for the most successful.[12] Despite the *Diary*'s name it was not a female-dominated world, though the degree of female participation by Hutton's time is badly obscured by the prevalence of pseudonyms among the contributors. Indeed, there was a sense in which the original project of making mathematics a subject of polite discussion had failed by this time, as the *Diary* and its siblings succumbed to pressure for harder, more intellectually competitive problems; it had become, in part, a world in which ambitious mathematical practitioners displayed their prowess.[13]

Hutton contributed to the *Ladies' Diary*, the *Gentleman's Diary*, and *Martin's Magazine of the Sciences* during the 1760s, as did many other northerners from his circle.[14] For a spectrum of different contributors—schoolmasters, schoolchildren, tradesmen, leisured enthusiasts—these periodicals were a site in which a degree of prestige and even, in a limited sense, fame could be acquired, for the investment of a few hours or tens of hours of work on sometimes difficult, abstruse, or laborious mathematical problems. The philomath periodicals also printed predictions of solar and lunar eclipses, and these too were a way to demonstrate, indeed to advertise, saleable skills. So for some—including Hutton—the philomath world was one to exploit as well as to enjoy.

Hutton's first major exercise in self-promotion, indeed, quite probably exploited the philomath network. In 1768–70 he produced his *Treatise of Mensuration*, a sort of encyclopaedia of geometry, boldly intended in effect as a replacement for Euclid's *Elements*. It was published by subscription and issued in monthly instalments, and Hutton evidently devoted much effort to soliciting subscriptions.[15] When the *Mensuration* appeared as a complete volume in 1770, more than 600 subscribers were listed. Geographically they ranged from Penzance to Dundee, socially from the nobility to the tradesman; all the English and Scottish universities and a number of other prominent institutions were represented. Many of the individual subscribers were schoolmasters: some excisemen or architects.[16] We do not know for certain how Hutton assembled this list of supporters; some will certainly have come through public advertisements but later evidence shows Hutton writing personally to friends and acquaintances to encourage purchase of his works or subscription to them, sometimes enclosing printed single-leaf advertisements.[17] I believe that in this case he did the same, and that the philomath journals—surely—provided him with many of the names to whom he wrote. Whatever the exact mechanism, the result was that Hutton's name was now very firmly on the philomaths' map as well as the practitioners'.

Woolwich and Greenwich

His local network of patronage and support eventually helped Hutton to move away from Newcastle. In 1773 an open competition was announced for the professorship of mathematics at the Royal Military Academy in Woolwich. Despite promises that the competition would be impartial and based on merit alone, the candidates armed themselves with the usual array of recommendations and testimonials; Hutton did the same. He travelled to London bearing recommendations from the Duke of Northumberland, the Earl of Sandwich, and the eccentric northern mathematician William Emerson, and it was later said that the Duke interfered in the process of the competition on his behalf. Through a combination of merit and interest, then, Hutton—having duly impressed the judges—was appointed as the new professor at Woolwich.[18]

The Royal Military Academy existed to train the cadets of the Royal Regiment of Artillery, and it had undergone a series of changes to its structure over the previous few years which had produced resentment among the staff. Merely by doing his duty and not indulging in open rebellion against his superiors Hutton was able to ingratiate himself with them, and over the years following his appointment a series of adjustments to the curriculum increased both the prominence of mathematics and Hutton's control over cadets' progress through the Academy; he became *de facto* the principal member of the academic teaching staff as well as the secretary of a 'Military Society' based at Woolwich which existed to pursue experimental work on military subjects.[19]

Several testimonies indicate that Hutton was liked and admired. His teaching issued forth in an account of conic sections and, later, a comprehensive textbook of mathematics.[20] Under him there was established an elaborate system of separate graded classes, with several assistants to teach them. Within a few years of Hutton's retirement all this would be the object of strident criticism, and it seems that things may have ossified towards the end of Hutton's time at the Academy, with the students doing little more than mechanically cramming his printed *Course* and its exercises.[21] Yet at the time of Hutton's death in 1823 a number of commentators remarked on the benefits he had brought to military training and tactics, and the Duke of Wellington paid public tribute to his work; no-one would have known better what value three generations of Hutton-trained gunners and engineers had brought to various theatres of war over the previous fifty years.[22]

After his move to Woolwich Hutton became part of another local network of mathematical practitioners, this time centred on the Royal Observatory at Greenwich. He wrote in 1779 of being 'almost as recluse as a hermit', but, as letters to observatory assistant Reuben Burrow and Astronomer Royal Nevil Maskelyne show, he was in fact quick to make links with the next village up the Thames, borrowing and lending books, for instance.[23] Maskelyne had been one of the examiners for his new position at Woolwich, and their letters illustrate the continuance of a relationship that had something of tutelage about it, Hutton asking Maskelyne's advice about the presentation of his mathematical ideas with a view not just to their correctness but to their attractiveness to readers ('I am apprehensive the introduction to the Geometrical paper might be better expressed').[24] The correspondence of David Kinnebrook (assistant at the Royal Observatory from 1794 to 1796) also bears witness to Hutton's relationship with the Observatory and its staff; there continued, too, to be some traffic of personnel between the Observatory and the Academy.[25]

Hutton also remained in contact with the national philomath network, not as a contributor now but as its controller. Soon after his arrival in Woolwich, Hutton was asked to succeed the previous editor of the *Ladies' Diary*, Edward Rollinson, who had died earlier in 1773. He took charge of certain other almanacs too, and by 1786 he was receiving 130 guineas a year to compile nearly all the annual almanacs produced by the Stationers' Company of London.[26] The *Gentleman's Diary* was kept somewhat independent from the *Ladies'*, but when its editor Thomas Peat died it was Hutton who 'procured' the editorship for Charles Wildbore, who had himself come to his attention through the philomath network.[27]

Meanwhile, Maskelyne soon admitted Hutton to a more exclusive mathematical network that he personally controlled: the team of computers and comparers of the *Nautical Almanac*. Maskelyne's coordination of a national team of calculators to make the lunar tables that were his solution to the 'longitude problem' is fairly well known.[28] The group overlapped with the philomath world and with the mathematical-practitioner world—nearly all the computers were one or both—but it was distinct from either group, and operated by distinctive rules. There were no prizes for doing better than the other computers; there was no publicity. Computers were never named in the printed tables but they were paid for their work, and they could be punished for doing it badly.

Hutton appears in Maskelyne's accounts for the *Nautical Almanac* during 1777–9; he worked as 'comparer', checking and revising the work of the computers, for a total of 12 months' worth of published almanacs.[29] We do not know whether this had the character of a favour Hutton did for Maskelyne—filling a gap in the team for the year—or a test of Hutton's abilities as a calculator. Perhaps it was both.

Via the *Almanac* Hutton became visible to another important institution, the Board of Longitude. By 1779 Hutton was cultivating direct contact with the Board, doing odd pieces of work such as correcting the proofs for a mathematical book and translating its preface.[30] He also put together a set of tables of products and powers of numbers, presented it to the Board and was duly rewarded: the Board paid for its printing.[31] The success of this volume and its adoption as part of the equipment issued to computers on the *Nautical Almanac* led Hutton to undertake a second, larger volume, of logarithm tables. This time Maskelyne supplied two short portions of text incorporated in the preface, and received the volume's dedication in recognition of the 'generous advice and assistance' he had given to this work and others.[32] The *Tables* appeared in 1785 and became one of Hutton's longest-running books, still in print in 1894.

Maskelyne was also a member of a still more prestigious network. He was a Fellow of the Royal Society; indeed, a member of its Council. The Royal Society retained in this period a gentlemanly ethos of disinterestedness as well as an increasing focus on public utility, and it was difficult—it was known to be difficult—for mathematicians to gain a hearing there.[33] They were often interested—in the financial sense—in the results of their investigations, and their work, when it was not merely commercial in orientation, had a habit of being incomprehensible. Maskelyne's foe John Harrison, indeed, was a recent example of a mathematical practitioner who had notably failed to put aside his commercial values in dealing with the Board of Longitude and the Royal Society.[34] It was, furthermore, thought by some that Fellowship of the Royal Society required certain social qualifications. Hutton's education—not perhaps particularly unusual in itself—rendered ambiguous his relationship to the shifting ground of Georgian social status and social mobility, and while some accepted him as a gentleman others would continue to the end of his life to insist he was none. Nevertheless, he was successfully proposed for Fellowship of the Society in 1774, by a list of signatories which, naturally, included Nevil Maskelyne.[35]

Near the end of his life—and, perhaps significantly, after most of the other protagonists were dead—Hutton recalled this period at the Royal Society in affectionate terms:

> This situation [fellowship of the Royal Society] brought me into immediate and intimate habits of familiar acquaintance with most of the more scientific characters of that respectable body, and about the Metropolis. Of those there was a small select number of gentlemen who constituted themselves into a certain club, that, after the usual weekly meeting of the Royal Society, retired to Slaughter's Coffee-house, at the top of St. Martin's Lane, to eat a few oysters, and hold familiar di[s]course together on the subjects that had occurred at the Society's meeting, and on any other current scientific matters. I constantly attended this club as an adopted member[.][36]

Hutton named among the members of what was later called the Royal Society Club Constantine Lord Mulgrave, Joseph Banks, Daniel Solander, Charles Blagden, George Fordyce, and John Hunter. This was the most exclusive network British natural philosophy had to offer, and Hutton was for a period its golden boy. In five years he published six papers in the *Philosophical Transactions* on subjects ranging from the solution of equations to the density of the earth. The latter arose from experimental work by Maskelyne (who else?) at Schiehallion in Scotland; Hutton had been commissioned to undertake the exhausting calculations to assist with their interpretation. Maskelyne received the Society's Copley Medal in 1775 for the project.[37] In 1778 Hutton himself received the medal for his ongoing experimental work at Woolwich on the explosive force of gunpowder, indicating decisively that he now had an identity not just as a mathematical practitioner, teacher, and calculating assistant but also as an experimental natural philosopher.

Thereafter, however, things began to unravel for Charles Hutton and, as I have argued elsewhere, what went wrong can be seen as a failure in Hutton's otherwise largely successful career as a worker of networks.[38] So far he had been conspicuously adept at using his connections in one world to gain entry into another: from teacher to philomath, from philomath to author, from author to professor of mathematics, and so on. More could be said about this, although the limitations of our evidence do not always enable us to see clearly how Hutton operated. In about 1779, however, it became clear that some of the circles in which Hutton moved—some of the associations he carried with him—were toxic to some of the people with whom he now wished to associate.

Specifically, by 1779 his friendship with the (now ex-) President of the Royal Society, John Pringle, was itself a liability, Pringle having fallen from royal favour due to his support for Benjamin Franklin; his circle also included the increasingly problematic radical Joseph Priestley.[39] For neither

Franklin nor Priestley do we have direct evidence of an acquaintance with Hutton but it is surely significant that again and again the colleagues and associates named in Hutton's surviving correspondence are nonconformists, political radicals, or both: Charles Babbage, Francis Baily, John Bruce, David Steuart Erskine, William Frend, Olinthus Gilbert Gregory, James Hutton, John Leslie, Samuel Lowthian, Paul Henry Maty, Alexander Tilloch, William Turner, George Walker. Hutton himself had been a Methodist in his youth, had attended the unitarian chapel at Hanover Square in Newcastle, and expressed in private at least a degree of enthusiasm for American liberty.[40] These connections were evidently enough to taint him.

Hutton therefore brought to Fellowship of the Royal Society not just a willingness and ability to socialize over oysters and coffee, and not just high-level abilities as a mathematician and expositor. He also brought palpable connections with the worlds of radical politics and religious dissent, as well as the worlds of trade, teaching, and metropolitan journalism. To all of these the new president of the Royal Society, Joseph Banks—indeed any president of the Royal Society in the political situation around 1780—could only take exception.[41] At the end of 1779 Hutton had become a member of the Royal Society's council and its Foreign Secretary; by the end of 1783 Banks had seen to it that he was no longer either. The ensuing row held up the scientific business of the Society for several weeks early in 1784 and came close to unseating Banks as President, but the sheer number of Banks's supporters within the Royal Society meant that there was nothing Hutton's friends, disparate as they were, could do, and in the event it was he and a number of others who went into self-imposed exile from the Society, ceasing to publish in the *Philosophical Transactions* and absenting themselves from meetings.[42]

London and beyond

Hutton's life so far had seen a series of moves to increasing public visibility and institutional recognition; now that that process had stalled, he turned his attention at least in part to the prestige and visibility of his discipline, doing all he could to promote mathematics in the public sphere. In this context Hutton's various networks—including his remaining circle of friends within the Royal Society—came to serve transformed and urgent purposes. The mathematicians of the Royal Society quickly formed a 'club' of their own, sometimes called the 'Friday dining club'; fortnightly meetings at the Mitre Tavern (probably that in Holborn) and subsequently the Globe Tavern in Fleet Street continued at least until 1802.[43] This was an intellectual support network without which the animosity of the President of the Royal Society would have placed Hutton's career in much greater jeopardy. There was some talk of producing a periodical publication to rival the *Philosophical Transactions*, but nothing came of that;[44] publications appeared in 1786 and 1812 under the guise of collections of 'tracts' by Hutton, though as Letter 31 witnesses, there was at least a small element of collaboration in their production. Equally, Francis Maseres' massive compendium, the *Scriptores Logarithmici*, incorporated work by Hutton and was sometimes thought to be a joint effort.[45] Hutton also continued as editor of the *Ladies' Diary*, producing in addition an annual *Supplement* at his own expense as well as the *Diary* proper, so as to put into print more of the philomathic material he was receiving. He also did a certain amount of work re-editing the work of deceased colleagues, another way of asserting his control and his importance in a community of British mathematicians.[46] Members of that community, meanwhile, would continue to campaign against Banks and his circle into the 1820s.

Ostensibly focused largely on the public utility of mathematics, Hutton's activities also had the effect of adding considerably to his own visibility, and some of his publications seem aimed as much as anything at consolidating his own reputation and establishing his own voice as the leading one speaking for mathematics in English. He produced his famous *Dictionary* in 1795–6, he gave his name to a translation of Ozanam's popular *Mathematical Recreations* in 1803 (I have suggested elsewhere that he did not in fact undertake the translation himself),[47] and he oversaw an eighteen-volume abridgement of the *Philosophical Transactions* published in 1809, stamping—and not by accident—the literary output of the Royal Society with his own interpretation through commentary and notes.

Throughout all this, Hutton also remained in touch with his northern circle of colleagues and friends. In 1779 his Edinburgh supporters had arranged the award of an honorary doctorate of laws; in 1795 and 1805 he wrote in support of John Leslie's candidacy for positions at that university.[48] The founding and progress of the Newcastle Literary and Philosophical Society attracted Hutton's interest and seems to have stimulated correspondence with a new generation of northern literati, some of whom visited him in London (although Hutton's plan to revisit the north-east was never realized): John and Edward Bruce, William Armstrong, and John Swinburne.[49] Hutton was elected to the Royal Society of Edinburgh as well as to the Philosophical Societies of Harlem and of America. Evidently still a clubbable man, he also participated in the Society of Civil Engineers and the Chapter Coffee House Philosophical Society.[50]

As the editor of the *Ladies' Diary*, Hutton was now, indeed, a dispenser of mathematical patronage on a substantial scale. As the Royal Military Academy grew under the pressure of the French wars, Hutton was able to add more and more mathematical staff; by 1807 he had six assistants. In all, eight mathematicians were brought to Woolwich as Hutton's assistants during his time there, all with his approval and nearly all at his specific recommendation.[51] The case of Lewis Evans is one we can document particularly fully through the chance survival of his side of his correspondence with Hutton, from the initial letters of the early 1780s addressed to the anonymous editor of the *Ladies' Diary*, through warmer, more personal exchanges in the 1790s including the gift of hares and pheasants. In 1799 he was employed at Woolwich as Mathematical Master under Hutton. A notable incident is recorded in the pair of letters from September 1788, when Evans revealed his panicked concern not to lose Hutton's approval after a rival teacher accused him of disparaging the older man's arithmetic textbook.[52]

When there was not space at Woolwich for all those Hutton wanted to favour, there were other institutions ready to attend to his endorsements. An obituarist stated that 'To his recommendations, as well as to his instructions, our most eminent scientific institutions have been chiefly indebted for their Professors of Mathematics, during the last thirty years'.[53] Examples mentioned within the correspondence include Edward Riddle and Margaret Bryan, both notable recipients of Hutton's encouragement and practical help.[54] Furthermore, both the East India Company's military seminary at Addiscombe and the Royal Military College at Sandhurst were set up with the Royal Military Academy at least partly in mind as models, and the mathematics teaching in both schools used Hutton's *Course* and naturally looked to him for a lead; at Addiscombe and at Woolwich he was also an examiner until about 1817.[55] Hutton had become the centre of a mathematical network of his own.

International contacts are very much less visible in the surviving correspondence, unfortunately making it hard to estimate how much personal contact Hutton had with colleagues on—in particular—the European continent: how far he participated in any genuinely international community of mathematicians. Hutton lived through a period when Continental mathematical methods were increasing in visibility in Britain.[56] Unlike some of his contemporaries and successors he did not hold out stridently for the superiority of fluxions over continental analysis, of Newtonian 'dot' notation over Leibizian 'd' notation. Certainly he admired his foreign colleagues; his publications regularly cited the works of Euler, Lagrange, and other continental luminaries with approbation, and he was appointed to the ill-fated

Foreign Secretary position at the Royal Society on the understanding both that he could read and write French (in fact he also read Italian and Latin, and probably German) and that he was able and willing to exchange pleasantries and flattery with foreign members.[57] Some time before October 1781 he was visited by Edme-Sébastien Jeaurat from the Paris Observatory; there were possibly other foreign visitors of whom we now have no record. Hutton's youngest daughter, Charlotte, and probably her sister, Eleanor, were educated at a convent school in France for a period.[58]

Hutton's library contained numerous foreign mathematical works of the eighteenth century, from d'Alembert to Varignon: over forty authors were represented by over eighty volumes, plus more than fifty volumes of foreign periodicals. He had multiple volumes of such advanced writers as Clairaut, Euler, Lalande, and Laplace. It seems likely that it was the best collection of continental mathematics in Britain.[59] His letter to Laplace of 1819 may be unique in Hutton's surviving correspondence, but the flattering view it took of continental science was certainly sincere, and the kindness of Laplace's reply stood in marked contrast to the attitude of Joseph Banks to Hutton and his kind.[60] Yet, as a pedagogue, Hutton had no option but to teach the methods and the notations the English understood, and the same was true of his popular writing and his original research, which unfailingly used the 'dot' notation, and which tended to gesture towards continental works without giving much detail about their contents. When calls for the reform of British mathematics came, from 1802 onwards, Hutton was therefore—and perhaps unfairly—identified by some as a representative of a mathematical culture that deserved to be swept away.[61]

Nevertheless, Hutton's efforts to maintain his reputation, and his efforts on behalf of mathematics, worked. There was a final reconciliation with the Royal Society after the death of Joseph Banks in 1820; Hutton was elected once more onto the Council of the Society and published one final paper in the *Transactions*. A marble bust was commissioned by subscription and presented to him at his home.[62] By the time of his death in 1823 he was widely acknowledged to be a major figure in British science, and he had done as much as anyone of his generation to increase the status and visibility of mathematics in the English-speaking world.

There can be no simple summary. Hutton participated in local and national networks of mathematical practitioners and philomaths, and in such specialist networks as Maskelyne's team of computers and comparers, as well as being a member of several formal societies such as the Royal Society and looser groupings such as the Friday dining club. Neither his relationships nor his letters can be placed in neat categories. What is perfectly clear, however, is that Hutton's rise to the positions he occupied, his education both mathematically and socially, and his ability to survive a serious fall from grace depended crucially on his membership of and his dexterity in exploiting a range of different networks: his ability to translate connections in one world into practical gains in status or income in another. He was no self-made man. Who he was and what he was were made and sustained by the personal relationships witnessed in this correspondence.

Hutton's intellectual legacy

Divergent interpretations of Charles Hutton already existed during his lifetime: the adulation of his friends and colleagues, the disdain of political adversaries; the equivocal praise issued by younger mathematicians more interested in Continental innovations than in the British tradition. Much the same can be said of posthumous interpretations of Hutton and his achievements, which were naturally shaped by his multiple personal networks.[63] His London friends, ignorant of the details of his northern background, presented obituaries that emphasized his roles in mainly metropolitan networks of natural philosophers, mathematicians, and teachers. Olinthus Gregory, Hutton's successor both at Woolwich and at the Stationers' Company, provided a glowing testimonial to his character, his habits, and his teaching practices, and provided much information that was reused by other obituarists, although his anticipated full-length biography never appeared.[64] Meanwhile, in the north a handful of admirers collected what personal testimony they could about Hutton's youth. John Bruce put together an account of his life for the Newcastle Literary and Philosophical Society that combined detailed information about Hutton's time in the coal pits with a less sure account of his career at the Royal Society.[65]

New editions of his books would keep Hutton's name before the public well into the second half of the nineteenth century, achieving a longevity which would have surely gratified their author. His early textbook on arithmetic was last printed in English—in part—in 1867, and was translated into Japanese in the 1870s. His *Tables* remained in use over a similar period. His *Course* written for the Royal Military Academy was last printed in 1860; it was used at West Point, at Sandhurst, and at the East India Company's school at Addiscombe until at least the 1830s.[66] The Royal Bengal Artillery seem to have been particularly keen on it, and between 1828 and 1840 translations into Gujarati, Arabic, Sanskrit, and Marat'ha appeared at Bombay and Calcutta. Hutton thus achieved an international reach matched by only a very few of the mathematical authors of his period.

On the other hand, there was no clearly definable mathematical innovation to link with his name, and he was strongly associated with a style of mathematics—fluxions—and with a mathematical world—that of the philomaths—that were falling rapidly out of fashion in the nineteenth century. The challenge of a disorganized *nachlass* combined with these factors to effect the waning of his once considerable fame within a generation or so.

Hutton's manuscript legacy

Hutton was a public figure for nearly fifty years, and he held positions such as editor of the *Ladies' Diary* and overseer of the Stationer's Company's almanacs that alone involved many hundreds of letters. It is unlikely that his total correspondence numbered fewer than 10,000 items. He was the author of a dozen scientific papers and a dozen books, several of which were reprinted and revised multiple times. He kept a personal journal during at least some periods of his life.[67] All of this must have generated a personal archive of some size.

That archive was affected, however, by several rounds of dispersal and destruction, beginning during Hutton's lifetime. His protegé Olinthus Gregory stated, for example, that the manuscript of Hutton's lectures on natural philosophy was destroyed in remarkable circumstances which he did not specify.[68] Some of the equipment used in Hutton's experiments on ballistics succumbed to a fire at the Royal Military Academy's Repository in 1802. Material may well have been left behind at the Academy when Hutton retired in 1807; the archive there was subsequently devastated by another fire, as was that at the Tower of London where Board of Ordnance papers were kept.[69] Hutton's library was dispersed by sale in 1816, apparently due to the continuing animosity of Joseph Banks; some manuscripts and scientific instruments were sold at the same time, and it is quite possible that other items were given away or sold during a period when Hutton was contemplating moving to a smaller establishment and felt that his scientific work was definitively over.[70] His contract with the Stationers' Company required the return of all relevant papers to the company when Hutton ceased to compile almanacs in 1818, but very little Hutton-related material survives today in the Company's archive.[71] Elsewhere, annotations on some surviving letters and manuscripts show that they were given to Olinthus Gregory, either during the latter part of Hutton's life or by his family after his death. The items that had reached Gregory were then dispersed, at least some of them by sale, after he died in 1841.[72]

Hutton's will left his London house and its contents to his daughter Isabella. She was quickly reconciled with Hutton's estranged grandson—her nephew—Charles Blacker Vignoles, and he moved into the house later in 1823.[73] Subsequently, after Isabella's death and the deaths of her siblings Henry and Eleanor, the bulk of the estate—worth several tens of thousands of pounds—seems to have passed to Vignoles. However, he stated in a letter of 1841 that 'Accident' had destroyed many of the papers of his family; he later reiterated that he had only a 'few scraps ... of the papers of my Grandfather'.[74] Today the Vignoles family papers contain only a small selection of Hutton's manuscripts, including just a handful of letters, evidently chosen for the evidence they give about Vignoles' early life; of other material from Hutton's lifetime they have nothing. (The antiquarian papers of Hutton's son Henry, now in the Advocates' Library in Edinburgh, do not seem to include any material from his father.) There may be a legitimate suspicion that 'accident' was a gloss for deliberate destruction of material that could have reflected badly on Vignoles himself or on Hutton, but the Vignoles papers do not seem to have been expurgated in other ways, despite containing a mass of Vignoles' own letters, such as to his wife, that do not always show him in a sympathetic light. It was suggested later that Henry Hutton had quashed any attempt to publish a biography of his father because he did not wish it known that he had worked in the coal pits; the story is plausible, certainly, but not verifiable.[75]

At the same time, some efforts were being made in other quarters to collect and preserve manuscripts from Hutton. The Newcastle Literary and Philosophical Society collected a few books from his library and acquired one letter; Augustus De Morgan also acquired a few books, as well as one manuscript and two letters, and they remain in the Senate House Library in London.[76] Another interested person—Joseph Clinton Robertson (1788–1852), founder and editor of the *Mechanic's Magazine*—acquired a small collection of Hutton's manuscripts from Gregory's sale, and they eventually reached Trinity College Cambridge.[77] Charles Babbage seems to have acquired some of Hutton's books; they passed into the Crawford Collection at the Royal Observatory in Edinburgh.[78] Some of the other material that had passed to Olinthus Gregory stayed at the Royal Military Academy and went subsequently to the museum (now closed) of the Royal Artillery.[79] Letters from Hutton naturally remained with the papers of their recipients in a number of cases. The trajectories of other items are less clear, and the present locations of many of the letters edited here from printed sources are not known. Hutton's surviving manuscript correspondence is, today, scattered among two dozen different archives. Most of what survives is material that had left Hutton's possession by the time of his death, and it is a far from random selection.

Thus, for example, correspondence with Hutton's family is conspicuous by its almost complete absence. So is correspondence relating to Hutton's work for the Stationers' Company, with the exception of a long series of letters from Lewis Evans, a contributor to the *Ladies' Diary*, which survive as drafts. Hutton's contact with scientific circles in Newcastle is better represented, as is his relationship with the Astronomer Royal Nevil Maskelyne and his network of assistants and calculators.

Most of the surviving letters contain no mathematics. They deal instead with personal and intellectual relationships, the publication of Hutton's books, and the manoeuvring of Hutton and others in pursuit of jobs, publication, reputation, and recognition for themselves and their discipline. To that extent they represent a window onto early modern mathematical culture of a kind that has rarely been set open before, and it is in this that their value and interest lie.

This edition

I have included not only personal letters in the narrowest sense but also published covering letters, dedicatory epistles, letters to the editor, and certain other public communications taking the form of letters. Thus, for instance, I have included the covering letters that introduced certain of Hutton's papers in the *Philosophical Transactions* as well as the covering letter with which he forwarded a 1786 paper of Edward Williams to the Royal Society of Edinburgh. I have included Lewis Evans' covering letters for his problems and solutions intended for the *Ladies' Diary*, though not the problems and solutions themselves; they are numerous and lengthy, and would have increased the length of this volume inordinately. (Nor have I included any other material printed in the *Ladies' Diary* or *Diary Companion*, much of which took the form of letters addressed to its editor.) I have included Hutton's 1820 correspondence with Laplace, which appeared in print in the *Philosophical Magazine* but partook to a degree of the personal as well as the public.

I have, however, excluded several (sometimes long) letters addressed to the editors of periodical publications which seem to me to fall clearly within the category of Hutton's publications rather than his correspondence.[80]

Lost letters are not reported, with the single exception of Letter 89, included because of the unusual interest of Hutton's continuing relationship with Joseph Banks and the Royal Society.

Letters between third parties that bear on Hutton's life and interests or otherwise mention him are not included (quantities of such letters exist among the Vignoles family papers, the correspondence of the staff of the Royal Observatory, and the correspondence of officers of the Royal Society including Joseph Banks). Three exceptions have been admitted: Letter 93 to the eleven-year-old Charles Blacker Vignoles, which seems clearly to have been intended for the eyes of his grandfather Hutton, and Letters 16 to Hutton's daughters and 133 to his son, both of which bear directly on Hutton's reputation and his relations with his contemporaries.

In total, 133 letters are edited, from the period 1770–1823, including 73 written by Hutton. The scattered state of his manuscripts makes it likely that more correspondence of Charles Hutton will come to light in the future, and I will be happy to receive notice of letters I have missed. The reader's indulgence is asked for the appearance of Letters 58 and 74 slightly out of their chronological sequence, a result of changes in the dating of certain letters at a late stage in the preparation of this volume.

~

Where the source is a printed text it is followed literally (with the exception of layout regularization as noted below), including any editorial interventions present in the source.

Where the source is a manuscript text, I give the author's final intentions, with authorial insertions distinguished by ⟨angle brackets⟩. Legible deletions and alterations are reported in footnotes. Illegible deletions are noted only where the sense of the text is impaired. Other editorial interventions—mainly indications of the loss of text or its conjectural replacement—are placed in square brackets.

The sender's address, if any, is given before the main text and the recipient's address, if any, after it. Indentation is regularized. Pagination is not recorded, neither are catchwords.

Headnotes to each letter indicate the source for the text and provide any physical details relevant to the edition, such as later endorsements when they give information not contained in the letter itself. They also record any previous printings of the letter. Endnotes to each letter identify correspondents and other people, publications and events where references made by Hutton and his correspondents might otherwise be obscure. Repetition is avoided—within reason—by the use of cross references.

A complete calendar of these letters is available at Early Modern Letters Online (emlo.bodleian.ox.ac.uk) and has been deposited at the Oxford Research Archive (ora.ox.ac.uk, DOI: 10.5287/bodleian:1aDGwMyxr).

Abbreviations

ALS: Autograph letter as sent.

DSB: Charles Coulston Gillispie, Frederic Lawrence Holmes, and Noretta Koertge, *Complete Dictionary of Scientific Biography* (Detroit, 2008).

ESTC: Electronic Short-Title Catalogue, www.estc.bl.

FR: 'FreeREG', www.freereg.org.uk.

IGI: International Genealogical Index, www.familysearch.org.

MPHE: E.G.R. Taylor, *The Mathematical Practitioners of Hanoverian England, 1714–1840* (Cambridge, 1966).

ODNB: *The Oxford Dictionary of National Biography* (Oxford, 2004).

Notes

[1]'Public Characters', 97–8; Bruce, *Memoir*, 2–3, 5; Gregory, 'Memoir', 201–2.

[2]Bruce, *Memoir*, 50; Woodhorn, SANT/BEQ/26/1/7/77 (notes of Thomas Wilson, 28 March 1822); Boyle, *Old Newcastle*, 152.

[3]Advertisements in the *Newcastle Courant*, 27 December 1766 and 25 April 1772; Hutton, *Tracts* (1812), vol. 3, p. 379; Hans, *New trends*, 110.

[4]Hutton, *Guide*; Hutton, *Mensuration*; Hutton, *Plan*; Hutton, *Diarian Miscellany*.

[5]Letter 2.

[6]Bruce, *Memoir*, 13; Mackenzie, *Historical Account*, 558; Welford, *Men of Mark*, *s.v.* George Coughron.

[7]Taylor, *Mathematical Practitioners*.

[8]Winterburn, 'Philomaths'.

[9]Woodhorn, SANT/BEQ/26/1/8/584: Letter of S. Barrass to Thomas Wilson, 1825.

[10]Gregory, 'Memoir', 202; Bruce, *Memoir*, 6–7.

[11]SANT/BEQ/26/1/8/584: Letter of S. Barrass to Thomas Wilson, 1825; Bruce, *Memoir*, 7, 10.

[12]Archibald, 'English Mathematical Serials'; Pedersen, 'The "Philomath"'; Costa, 'The *Ladies' Diary*: Society, gender and mathematics'.

[13]Perl, 'The Ladies' Diary'; Costa, 'The *Ladies' Diary*: Gender, mathematics, and civil society'.

[14]Martin, *Miscellaneous Correspondence*, vol. 4, pp. 767–8 (December 1761), and subsequently up to December 1763; see 'Public Characters', 103 for the identification of 'Tonthu' as Hutton; *The Gentleman's Diary* (London, 1763); *The Ladies' Diary* (London, 1764–73), *passim*.

[15]Proposal for printing the *Mensuration* by subscription, *Newcastle Chronicle*, 26 December 1767.

[16]Hutton, *Mensuration*, v–xiv; Wallis, 'The book subscription lists project', 323; Wallis and Wallis, *Mathematical tradition*, 36.

[17]Letters 70 and 96.

[18]'Memoir of the late Dr. Hutton', *Gentleman's Magazine* (March 1823), 228–32 at 229; 'Public Characters', 108; Woodhorn, SANT/BEQ/26/1/8/584: Letter of S. Barrass to Thomas Wilson, 1825; Bruce, *Memoir*, 15–17.

[19]Jones, *Records of the RMA*, *passim*; Museum of the Royal Artillery ('Firepower'), MD 913/5, item 2, fol. 2^r; Smith, *Dictionary*, *s.v.* 'Gunnery'; Duncan, *History*, 168–70; Johnson, 'Charles Hutton', 224.

[20]Gregory, 'Memoir', 219–20; Fullom, *Douglas*, 16–17; Hutton, *Conic Sections*; Hutton, *Course*.

[21]Jones, *Records of the RMA*, *passim*; Saint, *Four Letters*.

[22]Obituaries of Hutton in the *Literary Chronicle* (1 February, 1823), 77, in the *London Magazine*, (March 1823), 368, and in the *Monthly Magazine* (March 1823), 142.

[23]Letters 4, 9 (quote), 13.

[24]Letter 31.

[25]Cambridge University Library, RGO 35/106, 35/108 and 35/115: letters of David Kinnebrook to his father, 23 February, 26 March and 29 October 1795.

[26]Myers, 'The Records of the Stationers' Company': Court book M, 483; Supplementary documents Series I, Box B, folder 6, items i–iv; Court book O, 77.

[27]Letter 14.

[28]Croarken, 'Longitude for All'; Croarken, 'Tabulating the heavens'.

[29]Cambridge University Library, RGO 4/325, fols 54^v–55^r.

[30]Cambridge University Library, RGO 14/5, pp. 344, 347–8 (minutes of Board of Longitude meetings of 6 March and 10 July 1779) and 14/7, pp. 337–9 (account of the Board of Longitude with Charles Hutton, 1779–82).

[31]Hutton, *Tables*; Cambridge, University Library, RGO 14/6, pp. 6–7 (minutes of Board of Longitude meeting of 15 July 1780).

[32]Croarken, 'Tabulating the heavens', 55; Hutton, *Tables*, 156, 168–9.

[33]Wardhaugh, 'Charles Hutton and the "Dissensions"'; Heilbron, 'A mathematicians' mutiny'.

[34]Higgitt, *Maskelyne*, 231.

[35]London, Royal Society EC/1774/18: proposal of Hutton for Fellowship of the Royal Society, read 10 March 1774.

[36]Hutton, 'To the Editor of the Newcastle Magazine', 127.

[37]Maskelyne, 'An Account'.

[38]Wardhaugh, 'Charles Hutton and the "Dissensions"'.

[39][Gregory], 'A Review', 170–72; Franklin, *Benjamin Franklin*, vol. 1, pp. 322–4.

[40]'Public Characters', 100; Mackenzie, *Historical Account*, 370–6; Letter 9.

[41]See also Watts, 'We want no authors'.

[42]But see Letter 28 and the note there.

[43]Cambridge University Library, RGO 4/187/11:1–2 (letter of Hutton to Maskelyne, 27 June 1785); MS Add. 7886/117 (letter of Hutton to William Frend, 21 May 1791); RGO 4/187/18 (letter of Hutton to Maskelyne, 19 December 1793); RGO 4/187/22: 22:1–2 (letter of Mr Rowed of the Globe Tavern, Fleet Street to Maskelyne, June 1795); British Library, Add. MS 37915, fols. 218–19 (letter of Hutton to Windham, 15 September 1802); see also Howse, *Nevil Maskelyne*, 161.

[44]Woolwich, Royal Artillery Museum, MS 913/5f (unattributed press clipping of 1784 or 1785); Chambers, *Correspondence of Banks*, vol. 3, pp. 92–4, 105–6, 110–11 (letters 605, 612, 614: Blagden to Banks, 30 September, 23 October, 30 October 1785).

[45]Letters 7 and 8; Letter 39.

[46]Simpson, *Select Exercises*; Whitehurst, *Works*; Robins, *New Principles*.

[47]Wardhaugh, *Gunpowder and Geometry*.

[48]Letter 12; Letters 73 and 95.

[49]Letters 86, 104, 106, 109, 115, 112, 113, 116, 118, 117, 130, and 129.

[50]Account of the newly formed Society of Civil Engineers in *Nicholson's Journal* 2 (1798), 47–8 and 95–6, at 96; Levere and Turner, *Discussing Chemistry and Steam*, 57.

[51]John Bonnycastle (1782), Lewis Evans (1799), Thomas Evans (1802), Olinthus Gregory (1803), Samuel Hunter Christie, Thomas Myers, Peter Barlow, William Moore (all 1806): Jones, *Records of the RMA*, iii–iv.

[52]Letters 42 and 43.

[53]Gregory, 'Memoir', 484; see also Bruce, *Memoir*, 37; *Tribute of Respect*, 2; [Gregory], 'Memoir of Charles Hutton', 226–7; and Letters 73, 74, 95, 114.

[54]See Letters 77 and 129.

[55]Vibart, *Addiscombe*, 5, 118; Jones, *Records of the RMA*, 47–8, 63; [Gregory], 'Memoir of Charles Hutton', 484; see also Swetz, 'Robert Adrain', 340.

[56]Guicciardini, *The development*; Guicciardini, 'Dot-Age'; Crosland and Smith, 'Transmission of Physics'; Craik, 'Mathematical Analysis and Physical Astronomy'.

[57]Kippis, *Observations*, 63, 74.

[58]See Letter 16 and notes there.

[59]*A Catalogue*, *passim*; see Letters 115 and 129.

[60]Letters 120 and 122.

[61]See Wardhaugh, *Gunpowder and Geometry*.

[62]Royal Society, JBO/43, 522 (30 November 1822); Hutton, 'On the Mean Density'; *Tribute of Respect*.

[63]Wardhaugh, 'One of the Greatest'.

[64]Gregory, 'Memoir'; [Gregory], 'Memoir of Charles Hutton', esp. 220; see Bruce, *Memoir*, 1.

[65]Woodhorn, SANT/BEQ/26/1/7/77 (Thomas Wilson, notes on Mr Kirkley's reminiscences of Charles Hutton, 28 March 1822), SANT/BEQ/26/1/7/98/a (Thomas Wilson, notes including information about Charles Hutton, 11 July 1824), SANT/BEQ/26/1/8/584 (Letter of S. Barrass to Thomas Wilson including George Parkin's reminiscences of Charles Hutton, 1825); also Birmingham Archives, Heritage and Photography Service, MS 3597/103/3 (title page for a projected 'Collections relative to Charles Hutton'); Bruce, *Memoir*.

[66]See note 55, Howson, *Mathematics Education*, 251, and Guicciardini, *The development*, 110.

[67]Gregory, 'Memoir', 220–1.

[68]Gregory, 'Memoir', 222.

[69]Museum of Artillery, *Catalogue*, vol. 1, p. 5; Jones, *Records of the RMA*, xiii, 115.

[70]See Letter 115 and *A Catalogue*, title page.

[71]Myers, 'The Records of the Stationers' Company': Supplementary documents Series I, Box B, folder 6, items i–iv.

[72]Southgate and Son, *Catalogue*; endorsements on Letters 19 and 72 and on Trinity College, Cambridge, MS R.1.59.

[73]The National Archives, PROB 11/1666/375, fol. 297^{v}: Will of Charles Hutton; Portsmouth, History Centre, Vignoles papers, Letters 271–84.

[74]Portsmouth, History Centre, Vignoles Papers, Letter 751: Charles Blacker Vignoles to the Marquis of Northampton, 25 March 1841; London, University College archives, MS Galton 2/4/1/2/9: letter of Charles Blacker Vignoles to Francis Galton, 17 November 1865.

[75]Bruce, *Hand-Book*, 153.

[76]The letters were nos. 102 and 103, the manuscript Senate House Library MS 235, a translation from Tartaglia by Hutton.

[77]Trinity College, Cambridge, MS R.1.59.

[78]Forbes, 'The Crawford Collection', 459.

[79]Woolwich, Royal Artillery Museum ('Firepower'), MD/913, a miscellaneous collection of papers of Charles Hutton and Olinthus Gregory.

[80][Rules for cutting timber], *Newcastle Courant* (13 February 1768); 'To the Editor of the Monthly Review', *Monthly Review* 38 (June 1802); 'To the Editor of the Monthly Magazine', *Monthly Magazine* (August 1802); 'To the Editor', *Monthly Magazine* (July 1820) (although signed 'Vindex' and dated from Blackheath this was reprinted in the *Philosophical Magazine* (no. 266, June 1820, pp. 465–8), in the *National Recorder* (9 September 1820, pp. 161–2) and in the *Newcastle Magazine* (no. 2, January 1823, pp. 48–50) as a 'Note on the Divining Rod' by Charles Hutton); 'To the Editor of the Newcastle Magazine', *The Newcastle Magazine* (March 1822).

LETTERS

Letter 1.
Charles Hutton to Hugh Percy, 1770

Dedicatory epistle printed in Hutton, *Mensuration*, iii–iv.

TO
HUGH PERCY,
Duke and Earl of
Northumberland.

My Lord,

I AM well aware how presumptive the world will think me in being desirous of having the name of one of Your illustrious rank, and universally confessed abilities in every branch of literature, prefixed to the performance which I now have the honour to present to Your Grace: Yet I hope it will do me justice, and not think that I am weak enough, in so doing, to have the vanity to suppose I had any thing to offer in this work unknown to Your Grace. But if the work have any merit in itself, and may be of any use to those less conversant in the liberal arts and sciences than Your Grace, my intention, I hope, will fully apologize for my ambition. For Your name, I am convinced, will attract the attention of many, who might not other wise look into the work; and should such receive any benefit or advantage from the perusal of it, I am sure, Your Grace will feel that satisfaction which none but the *real* Patrons of the scientific arts, and who have the *real* Interest and Love of their Country *truly* at heart, can feel.

To acknowledge Obligations which it is an Honour to have received, would still carry with it the appearance of Greater Ambition; and an attempt to describe those Virtues and Perfections which Your Grace's Public Works and Public Services have rendered so conspicuous, would be a Weakness indeed!

Permit me, then, humbly to assure Your Grace, that I have and shall ever retain the most grateful sense of the Favours and Honour done me; and that I am,

With the profoundest Respect,
YOUR GRACE'S
Most obedient and
Most devoted humble Servant,
CHARLES HUTTON.

Notes

Hugh Percy (1712–86), politician, had been Earl of Northumberland since 1750 and Duke since 1766: ODNB. The family's property in the north-east included coal mines, while the Sidegate in which Hutton was born was subsequently renamed Percy Street (its present name) after Hugh Percy. This and Letter 38 are the main evidence for Hutton's relationship with the Percy family; according to the recollection of one George Parkin, Hugh Percy both 'encouraged' Hutton to apply for the post at Woolwich in 1773 and interfered in the process of the election, securing after the first day a fresh set of questions 'that every one might have a fair chance': Woodhorn, SANT/BEQ/26/1/8/584 (letter of S. Barrass to Thomas Wilson including George Parkin's reminiscences of Charles Hutton, 1825).

Letter 2.
JOHN RICHMOND to CHARLES HUTTON, 30 July 1770

East Sussex Record Office, SAS/RF 18/141, 1^{v}: copy.

Newcastle 30th. July 1770.

Sir

As the Plan within alluded to was early communicated to you, and then met with your approbation; You'l please to inform me, whether you still entertain a favourable opinion of it. I am
Sir
Your mo: hble $Serv^{t}$.
Jno. Richmond

To
Mr Charles Hutton

Notes

John Richmond (b. 1740) was a Newcastle-based inventor and perhaps engineer; he subscribed to three mathematical books including Hutton's *Mensuration* and was still active in 1784: Wallis and Wallis, *Index*, 114. This letter and its successor are parts of a set of fair copy letters concerned with Richmond's new method for gauging the weight of 'keels': small, flat-bottomed vessels used for loading colliers. During late July and early August 1770 he wrote successively to John Rotheram, Charles Hutton, Robert Harrison, William Emerson, and William Brown; each replied favourably.

Letter 3.
CHARLES HUTTON to JOHN RICHMOND, 30 July 1770

East Sussex Record Office, SAS/RF 18/141, 1^{v}: copy.

Newcastle 30th July 1770.

Sir

The Sentiments above expressed by D^{r}. Rotheram[1] concerning the Scheme in question, perfectly agree with those which I always did and do still form of it — I am Sir

Your most Obed^{t}. Servt.

Chas. Hutton.

To
M^{r}. Richmond

Notes

For John Richmond see Letter 2.

[1]Probably John Rotheram, MD (*c.* 1719–87), who moved to Newcastle in 1760 as physician of the infirmary there: ODNB *s.v.* 'John Rotheram (*c.* 1750–1804)'. (Melmore, p. 73, confuses him with his son). He also gave public lectures, including some at Hutton's school in Westgate Street: Robinson, 'Trends in Education', 305. In a letter to John Richmond dated 28 July 1770 Rotheram wrote that

> I have with due attention, considered the plan you laid before me for weighing the Keels, and think I can undertake to demonstrate that it is fully adequate to the purposes, for which it is designed, that it will be easily carried into execution, and that it is plain and obvious to the meanest capacity, I am farther of opinion that the Vending of Coals by Weight, under this improvement, would be preferable, in point of certainty to any kind of measure.

Letter 4.
REUBEN BURROW to CHARLES HUTTON,
24 September 1773

UCL Library Services, Special Collections, MS Graves 23/3/5: ALS.
Date from endorsement on cover.

Royal observatory.

Dear Sir

I have sent home Wilsons Trigonometry[1] which you was so obliging as to Lend[a] me, and shall take the liberty to beg the favour of a sight of the "Miscellanea scientifica curiosa" publishd by M[r]. Wales[2] if you can spare it a few days as I want to see a paper on Occultations that is in it —

If I have any thing which you could like to see among my books you need only mention it & the old fellow shall bring it with him —

I am Sir
Your most humble Servant &c
Reuben Burrow

M[r]. Hutton
Professor of Geometry
at Woolwich
Academy

Notes

Reuben Burrow (1747–92), mathematician and orientalist, was at this time an assistant at the Royal Observatory at Greenwich. He later participated in the Schiehallion experiment (see Letter 120), and would become a bitter enemy of Charles Hutton thereafter, apparently because he felt he did not receive the credit that was his due. ODNB; MPHE; also Wilkinson, 'Mathematics and mathematicians'; and Senex, 'Board of ordnance in other days'.

[1]Either Henry Wilson (1673–1741), *Trigonometry improv'd* (London, 1720) or John Wilson (d. 1746), *Trigonometry* (Edinburgh, 1714, 1724).

[2]Six issues of a periodical titled *Miscellanea scientifica curiosa* appeared without imprint or author's name from 1766. They contained, according to the ESTC, 'articles on grammar, history, astrology, navigation, and mathematics; with puzzles, enigmas and mathematical problems' as well as correspondence. William Wales (1734–98), astronomer and mathematician, was one of two astronomers who sailed to Hudson's Bay to observe the transit of Venus in 1769 (see Letter 120): ODNB, MPHE.

[a] *altered from*: send.

Letter 5.
CHARLES HUTTON to SAMUEL HORSLEY, 25 January 1776

Printed in Hutton, 'A New and General Method', 476–7.
The paper which this letter accompanied was read to the Royal Society (in part) on 2 May 1776: London, Royal Society, JBO/28, p. 448.

Royal Mil. Acad.
Jan. 25, 1776.

SIR,

IN a late examination of the methods of Mr. MACHIN and others,[1] for computing the proportion of the diameter of a circle to its circumference, I discovered the method which is explained in the paper accompanying this letter. This method you, SIR, will perceive is very general, and discovers many series which are very fit for the abovementioned purpose. If you think it has sufficient merit to entitle it to the honour of being offered to the Royal Society, I have taken the liberty to inclose it to you, requesting the favour of you to present it accordingly, from, &c.

P.S. In the course of the last year, I attended some very interesting experiments on the effects of the force of fired gun-powder; as they seem to lead to many important conclusions, I believe I shall, in a short time, have the honour of submitting to your inspection an account of some of them, for the like purpose as the following paper.[2]

Notes

Samuel Horsley (1733–1806), bishop of St Asaph, was at this time a Secretary of the Royal Society. He would be a supporter of Hutton in the 'Dissensions' of 1783–4; he would also edit the works of Isaac Newton in 1779–85, using at Hutton's instigation the Newcastle-based engraver Ralph Beilby for the illustrations. The two later lost touch, and by 1805 (see Letter 92) Hutton could say that he did not know Horsley's London address. ODNB; MPHE; also 'Public Characters', 108; and Letters 9 and 14.

[1]The reference is to the series given at Jones, *Synopsis*, 243 and there attributed to the astronomer John Machin (?1686–1751); Machin's paper 'The Quadrature of the Circle' had also appeared as an appendix to Maseres, *Dissertation* in 1758 (for Francis Maseres and his relationship with Hutton see note to Letter 7).

[2]Hutton alluded to the experiments that would be described in Hutton, 'The Force of Fired Gun-Powder' in 1778.

Letter 6.
CHARLES HUTTON to SAMUEL HORSLEY, 17 April 1776

Printed in Hutton, 'A Demonstration of Two Theorems', 600.
This covering letter was read to the Royal Society on 27 June 1776, but the paper it accompanied was not: London, Royal Society, JBO/28, pp. 489–90.

Woolwich Warren, Royal Mil. Acad.
April 17, 1776.

REV. SIR,

THE following is a demonstration of the two theorems mentioned in Art. XXV. of the Philosophical Transactions for the year 1776.[1] If you think them proper to be inserted in the next volume of Transactions, be pleased to communicate them to the Royal Society for that purpose, from, &c.

Note

For Samuel Horsley see Letter 5.

[1]Lexel, 'Extract of a Letter from Mr. Lexel to Dr. Morton'.

Letter 7.
CHARLES HUTTON to FRANCIS MASERES, 13 March 1777

Printed in Maseres, *Scriptores Logarithmici*, vol. 6, p. 455.
Date from accompanying text.

Sir,

Having some leisure time at the receipt of your Letter desiring the Computation of the Sine of 1 Minute to 12, or 14, places of figures,[1] I have made that computation myself in the manner you wished to have it done, and I have herein inclosed you the whole process at large. Each equation is carried to 15 places of decimals, the last figure being always that which is nearest to the truth in the 15th place of decimals.—You will perceive that my manner of performing the Divisions, Multiplications, and Extractions of roots (which are strictly true,) saves a great deal of trouble in such operations.—I was always of opinion that Raphson's was an easier and readier Method of resolving high equations than Halley's, &c, as it is both clearer and requires less time.[2] But this Method of obtaining the Sine of $1'$ is far inferiour, in point of expedition, to that of Infinite Serieses. However, it is both quicker and more satisfactory than the old way, by the continual bisections of the arc of 30 degrees, as you very properly remark.

I remain, &c.

C. HUTTON.

Notes

Francis Maseres (1731–1824), colonial administrator and author, would publish from 1791 to 1807 *Scriptores Logarithmici*, a collection of writings about logarithms inspired by the lengthy historical introduction to Hutton's 1785 *Tables* and dedicated to Hutton: ODNB, DSB; on the *Scriptores Logarithmici* see also the references in Letters 39 and 67, which seem to indicate that the volumes were Maseres' work alone ('presented to' Hutton) but that some readers believed Hutton to have shared in the editorial work.

[1]Maseres had mentioned this desideratum to Hutton earlier in 1777: Maseres, *Scriptores Logarithmici*, vol. 6, p. 452. The method Maseres had evidently suggested is indicated in the full title the computation was given as printed following this letter, in the *Scriptores Logarithmici*: 'A computation of the length of the sine of a circular arc of one minute of a degree, by Mr. Raphson's method of resolving high equations by approximation; derived from the chord of an arch of sixty degrees, (which is equal to the radius of the circle,) by one quinquisection, one trisection, two bisections, a second quinquisection, a second trisection, and a third bisection, and by taking the half of the chord last-found.'

[2]The allusion is to the methods set out in Raphson, *Analysis aequationum universalis*, and Halley, 'Methodus Nova Accurata et Facilis Inveniendi Radices Æquationum Quarumcumque Generaliter, Sine Praevia Reductione'. Raphson's method is equivalent to that now called Newton's method or the Newton–Raphson method, but Raphson's 1690 description is differently expressed (it does not make use of the Taylor series).

Letter 8. CHARLES HUTTON to FRANCIS MASERES, 25 August 1777

Printed in Maseres, *Scriptores Logarithmici*, vol. 6, p. 456.
Date from accompanying text.

Sir,

You herewith receive again the Computation of the Sine of one Minute, with the proof of each of the seven extractions, or sections; which indeed took-up more time than I at first expected, the Labour being hereby nearly doubled. I have had a fair copy taken, in order to place each Proof on the right-hand page facing it's own proper section, as you will perceive.—You will perceive that I have availed myself of the Method you mentioned, of raising one power of a large number from [the same power of][a] another differing from it by an unit in the last place of figures;[1] and in these powers I have omitted such terms as I perceived would be too small to take into the general sum.

I am, &c.

CHARLES HUTTON.

Note

For Francis Maseres see Letter 7. The enclosure mentioned here was printed in Maseres, *Scriptores Logarithmici*, vol. 6, 457–74, following this letter.

[1]Maseres, in the accompanying text, referred to his *Tracts*, 317–20 for an illustration of this simple method in which—2.09455148^3 being known—2.09455149^3 is computed by expanding $(2.09455148 + 0.00000001)^3$ as a binomial.

Letter 9.
CHARLES HUTTON to ROBERT HARRISON, 13 January 1779

Printed in Melmore, 'Some Letters', 71–4.
Melmore reported (p. 71) that the originals of this and Letters 12, 14, 15, 67, and 71 were at the time of writing (1946) in the possession of the Rev. Angelo Raine of York. Their present location is not known.

WOOLWICH 13 *Janry* 1779.

MY DEAR SIR,

Nothing but the want of a proper occasion to trouble you could so long have prevented me from having the pleasure of addressing you. I hope you have enjoyed good health & every other happiness during the interval. As to my own health, it is pretty well; but I am here almost as recluse as a hermit, being almost single in my studies & manner of thinking in this place, my nearest neighbour in these respects being the Astron. Royal at Greenwich[1] with whom I have the honour to be on very good terms. I herewith send you a small assortment of *American* seeds some of them mixed with a little of that sacred soil so nourishing to the seeds of freedom! there is little more of

[a]Square brackets appear in the printed source.

it that we can now call ours! These seeds are the joint collection (chiefly in the neighbourhood of New York) of our friend Williams & a Cap. Blomefield who has just brought them over with him to be sent to the Durham hermit.[2] This latter gentleman (B) is a very intimate acquaintance of W's, & intends to do himself the pleasure to write to you soon, being furnished by W. with your address for that purpose. W. is now a full Captn. he having lately acceeded to a company, & would have come over at this time to Europe to join it but has been tempted to remain in A. by the temporary office of Major of Brigade which may be worth near 20 s. pr Day to him. He was become very fat & had his health well till lately that a dangerous fever seized him, from which he is just happily recovered. He sometimes I find snatches an opportunity, in spite of stern Bellona & the furies, of sacrificing to Minerva & the Muses, & by turns amusing himself & others with Mathematics, Music, Poetry, Painting, &c & diffusing rational entertainment all around him. If any plays or entertainments are to be represented, W. directs the machinery, paints the scenes, points out the subjects, teaches the Dramatis personae & the band of Music, writes and speaks the prologue & epilogue, acts the principal part &c, &c. He has lately been contriving or improving a circular sliding rule, has divided & made one himself, taught an Instrument Maker in N.Y. to execute them for his Brother Brigs. & has printed there the description & use of it. In short he is all in all.

I herewith also send you a few sets of three papers I have had printed in the Philos. Transactions, one of which brought me the honour of Sir Godfrey Copley's annual prize Medal, I have also sent as many Copies of the President's speech made at the presenting of the Medal.[3] After keeping a set yourself, let me request the favour of you to send a set to Mr. Emerson, to Mr. Geo. Anderson N.Castle, to Dr. Rotheram, Mr. Lowthian,[4] & such other persons as you may think proper together with my Compts to each gentleman. I wished to have had another paper of mine ready to have sent much larger than all the others, viz the Computations for determining the general attraction of matter, by which it appears that the mean density of the earth is to that of common stone, in the ratio of 9 to 5 nearly;[5] but as those sheets are not yet printed, they must be sent at another opportunity.

On the resignation of Dr. Horsley, one of the two reading & minuting Secretaries to the R.S. two candidates were put in nomination for his office, viz. the Rev. Mr. Maty (son of the late Dr. Maty many years secretary) & your hble servt.[6] It soon appeared so clear both to my friends & myself that the majority of the members usually attending the meetings seemed to give the preference to us in their wishes, that we contented ourselves with doing very little more then announcing my intention. My Competitor's party, on the contrary, went about, with every possible means of interest, to every member, & by one means or other surprized a great number into promises to vote for him, by which means it happened that we were left in a considerable minority by the votes of many against me whose wishes were for me. And with the declaration of all the Society that the office should be quite at my service the next opportunity if I continued in the mind to accept it. They also did me the honour to elect me into the Council for the ensuing year, & have presented me with what is sometimes called the Latin Secretaryship & sometimes the Foreign Secretaryship. The duty of it consists in translating into English such papers as are sent to the Society from abroad by Foreigners, those papers being usually written in French or Latin. They are translated in order to their being read in English to the mixed Company at the ordinary meetings of the Society. I have finished one very large translation already,[7] & as I have lately been not inattentive to such studies, I hope to acquit myself to satisfaction in the business.[8]

I mentioned Mr. Emerson, pray how does he do? I hope as well as can be expected for his age; present him with my best wishes for his happiness, & if you please mention to him my doubts concerning the truth of what he asserts in the Schol. near the bottom of pa. 433 of his Miscellanies, about Bernouilli's two probs, that he may consider what is proper to be done against a new edition if it be printed.[9] I can prove the truth of the conclusions of the two solutions in I think a manner that will satisfy him, & will communicate it to him if he think it worth his while, or if he be not himself already convinced of the truth of them. This conduct indeed concerning these two probs surprizes me much: for if he thought the solns wrong, he ought to have given true ones if they admit of them; but if not, the probs & false solns ought not to have come into his book: & his assertion that "the solutions are wrong, for both the quantities are infinite" is very extraordinary; for he well knows that the sum of a series is not always of an inf. magnitude when the number of terms is inf. & if he thought the sum inf. for any other reason, he should have shewn it. Possibly he does not yet know that we are not now to expect any more Fame's Palladins (perhaps you will say that is no loss), that terrible & ugly fellow who with such long strides pursued poor Tristram into France, having lately stretched the great Capt. Heath in terra.[10]

I am, Dear Sir, Yours most sincerely & affectionately,

CHA. HUTTON.

Notes

Robert Harrison (1714/15–1802), natural philosopher and linguist, taught at the Trinity House school in Newcastle from 1757: ODNB; MPHE; Melmore, 'Some Letters', 71.

[1]Nevil Maskelyne; see note to Letter 13. Hutton may have felt intellectually isolated but he was not literally 'as recluse as a hermit'; Margaret Ord had by this date borne his daughter Charlotte (see notes to Letters 31 and 67). His son Henry had entered the Royal Military Academy as a cadet in March 1776, and the evidence of Letter 16 suggests that his older daughters may also have been part of his household by this time.

[2]The 'Durham hermit' was possibly William Emerson: see note 4. Edward Williams (d. 1793) of the Royal Artillery was sometimes credited with persuading Hutton to apply for the professorship at the Royal Military Academy in 1773; he would later work on the Trigonometrical Survey: Gregory, 'Memoir', 206; Seymour and Andrews, *A History*, 12; Melmore, 'Some Letters', 72; he is mentioned in Letters, 12, 14, 15, and 20, and see in particular Letter 36.

Sir Thomas Blomefield, first baronet (1744–1822), was subsequently Inspector of Artillery and Inspector of the Brass Foundry at Woolwich Arsenal; with Hutton he oversaw the experiments on ballistics at Woolwich which resulted in several of Hutton's most celebrated publications: ODNB; Hutton, *Tracts* (1786), 104; Melmore, 'Some Letters', 72.

[3]The papers alluded to are Hutton, 'A New and General Method'; Hutton, 'A Demonstration of Two Theorems'; and Hutton, 'The Force of Fired Gun-Powder'; the Copley Medal was awarded for the last of these, and the President's speech was published as Pringle, *Theory of Gunnery*.

[4]William Emerson (1701–82), mathematician, lived in Hurworth (or Harworth) near Durham, and was noted for his personal eccentricities; according to Bruce, *Memoir*, 15–17, he provided a testimonial for Hutton's application to Woolwich in 1773: ODNB; MPHE; Hutton, *Dictionary*; see also Letters 12, 14, and 85.

George Anderson may well have been the master bricklayer and housebuilder of that name (*c.* 1726–98) who by the late 1740s was working from premises on Westgate Street in Newcastle, the same street as Hutton's final Newcastle school. By the 1780s he was notably wealthy: Welford, *Men of Mark*, vol. 1, p. 59; also Bruce, *Memoir*, 6; a 'Mr Anderson' is also mentioned in Letters 15 and 70.

For John Rotheram see note to Letter 3. Samuel Lowthian (d. 1780) was assistant minister at the chapel in Hanover Square, Newcastle from 1751, and chief minister from 1760 until his death; Hutton's daughters Isabella and Eleanor were baptized at that chapel. He also ran a school in Newcastle, where he had some reputation as a mathematician: Mackenzie, *Historical Account*, 374; Robinson, 'Trends in Education', 234; parish registers available via IGI and FR.

[5]Hutton, 'An Account of the Calculations'.

[6]For Samuel Horsley see Letter 5. Paul Henry Maty (1744–87), librarian, found himself impecunious—and was awarded the secretaryship of the Royal Society—when his anti-trinitarian views ended his clerical career: ODNB. Both men would be important players in the 'Dissensions' that erupted at the Royal Society in 1783–84, on which see Wardhaugh, 'Charles Hutton and the "Dissensions"'; the present passage provides unique evidence for Hutton's own view of his appointment as Foreign Secretary in 1779.

[7]This is most likely to have been Cazaud, 'Account of a new Method of cultivating the Sugar Cane', which was read to the Royal Society on 25 February 1779 and filled 80 pages when it was printed (in parallel English and French) in the *Philosophical Transactions*.

[8]By 1815 Hutton possessed books and dictionaries in Latin, Greek, French, German, Italian, Spanish, and Dutch (*A Catalogue*), but it is not certain with which languages except French and Latin he was acquainted at this time.

[9]Emerson, *Miscellanies* gave as Problem V:

'To find the value of

$$\sqrt{a+\sqrt{b+\sqrt{a+\sqrt{b}}}}$$

&c. ad infinitum.'

The solution printed there was in fact for the different expression

$$\sqrt{a+\sqrt{a+\sqrt{a+\sqrt{a}}}}$$

and was given as, correctly, $x = \frac{1}{2} + \sqrt{a + \frac{1}{4}}$.

Problem VI was:

'To find the value of

$$\sqrt{a\times\sqrt{b\times\sqrt{a\times\sqrt{b}}}}$$

&c. ad infinitum.'

Here Emerson gave, correctly, the solution $x = \sqrt[3]{aab}$. He then remarked in a scholium that 'These two Problems and their solutions are *J. Bernoulli's*; but the solutions are certainly wrong, for both the quantities are infinite', meaning I think that Bernoulli had printed incorrect solutions. Hutton mistakenly took the remark to mean that the solutions Emerson had printed were wrong.

[10]The allusion is to Robert Heath (d. 1779), editor of the *Ladies' Diary* in the 1740s and early 1750s and a supporter of Emerson: MPHE. From 1749 Heath supplemented the *Diary* with an 'appendix' separately titled *The Palladium*, and which continued as an independent publication after he lost the editorship of the *Diary* in 1753–54. Heath was a combative writer, and it is not clear which of the many objects of his attacks is meant by 'poor Tristram'.

Letter 10.
HENRY CAVENDISH to CHARLES HUTTON, 7 July 1779

Cambridge, Trinity College, MS R.1.59, fols 168, 170: ALS.

Text lost at several points due to damage.

Year given in later endorsement.

Marlborough street July 7

Sir

I am very much obliged to you for the trouble you have taken with my papers. I had no thoughts of desiring you to take so much trouble as to make the computati[on] you mention. My reason for wishing you to read the paper, besides the desire of having the opinion of a person of your judgement upon it in general, was that[a] though I believe the method containd in the first part is more convenient than that recommended by La Lande,[1] I am not sure that it is so; & as I thought you might very likely have turned your thoughts to that subject I was desirous of having your opinion of it. As to the latter part, I do not know that any one has given rules for determining the orbit of a comet in an ellipse, when the time of its revolution is unknown, except the proposition in Newton (inventam Cometæ trajectoriam corrigere)[2] which is very troublesome. If you know of any others I shall be glad if you would let me know

The 2 passages you take notice of are not errors in transcribing, but I believe are right. My meaning in the 1st is as follows

Let the mean anomaly of the point B be called A; & let the angle BSP & the distance SP both flow; the fluxion of the mean anomaly of B = $\frac{3P\times BS^2\times\dot{\overline{BSP}}}{2\sqrt{SP}} + \frac{3A\times \dot{SP}}{2SP}$, supposing $\dot{\overline{BSP}}$ to be given in parts of the radius; observing that by $\dot{\overline{BSP}}$ I mean the fluxion of the angle BSP

For A was before said to = $\frac{3P}{\sqrt{SP}}\times$ the area BSP; & the fluxion of the area BSP, supposing SP to remain constant, = $\frac{BS^2\times\dot{\overline{BSP}}}{2}$; & therefore if only BSP flowed, & SP was constant, the fluxion of A would equal $\frac{3P\times BS^2\times\dot{\overline{BSP}}}{2\sqrt{SP}}$. Moreover if SP flowed & BSP was con[stant,] the fluxion of A would equal $\frac{A\times 3\dot{SP}}{2SP}$; for if SP is increased or diminished in any ratio ⟨the angle BSP be[ing] constant⟩ A will be increased or diminished in the sesquiplicate ratio the[reof] & therefore if BSP & SP both flow the fluxion of $A = \frac{3P\times BS^2\times\dot{\overline{BSP}}}{\sqrt{SP}} + \frac{3A\times\dot{SP}}{2SP}$

The same rule may be demonstrated otherwise in this manner. Let the area BSP be called D; A was before said to equal $\frac{3P\times D}{\sqrt{SP}} = \frac{D}{SP^2}\times 3P\times SP^{\frac{3}{2}}$. Now the quantit[y] $\frac{D}{SP^2}$ is the same whatever is the value of SP; & therefore its fluxion is the same as it would be, if SP was constant & therefore = $\frac{BS^2\times\dot{\overline{BSP}}}{2SP^2}$; & the fluxion of $A = \frac{BS^2\times\dot{\overline{BSP}}\times 3P\times SP^{\frac{3}{2}}}{2SP^2} + \frac{D\times 3P\times 3\dot{SP}\times\sqrt{SP}}{2SP^2} = \frac{3P\times BS^2\times\dot{\overline{BSP}}}{2\sqrt{SP}} + \frac{3A\times\dot{SP}}{2SP}$

[a] *deleted*: was that.

The 2nd thing you take notice of goes on a supposition that e & t^2e are infinitely small in which case $\frac{1+t^2}{1+\frac{t^2e}{2-e}} = \frac{1+t^2}{1+\frac{t^2e}{2}} = \overline{1+t^2} \times \overline{1-\frac{t^2e}{2}}$

Your Obedient humble servant
H. Cavendish

To
Mr Charles Hutton
at Woolwich Warren

Notes

Henry Cavendish (1731–1810), natural philosopher, would later support Joseph Banks in the 'Dissensions' at the Royal Society, and Hutton would never accept his work on the density of the earth as superior to the Schiehallion experiment: ODNB; Jungnickel and McCormmach, *Cavendish*; also Letters 81 and 120. Cavendish had sent Hutton a copy of a paper 'On finding the Orbits of Comets in a Parabola from 3 Observations', which Hutton copied during June 1779: the copy is now fols 155r–165r in the same manuscript volume as this letter. Evidently Hutton made some remarks when he returned the original, to which this letter was a response; see further Letter 11.

[1]Parisian astronomer Joseph Jérôme Lefrançais de Lalande (1732–1807) (DSB) had published in his 1759 *Tables astronomiques* new tables for the position of Halley's comet and (pp. 70–129) an account of the methods by which they had been computed. The specific question of determining the orbit of a comet from three observations was dealt with at pp. 77–91.

[2]Newton, *Principia mathematica*, Book 3, Proposition XLII, Problem XXI.

Letter 11.
Henry Cavendish to Charles Hutton, ?July 1779

Cambridge, Trinity College, MS R.1.59, fol. 169: fragment of ALS.

S is the focus P the perihelion B the place of the comet

Let the area BSP be called D; & let the mean anomaly of B be called A. A was shewn in a former propos. to equal $\frac{3P\times D}{\sqrt{SP}}$ the fluxion of which is certainly equal to $\frac{3P\dot{D}}{\sqrt{SP}} - \frac{3PD\times \dot{SP}}{2SP^{3/2}} = \frac{3P\dot{D}}{\sqrt{SP}} - \frac{A\times \dot{SP}}{2SP}$

But it must be observed that $\dot{D}$ is not equal to $\frac{BS^2\times B\dot{S}P}{2}$ unless SP is constant. If BSP & SP both flow, the value of $\dot{D}$ may thus be found. The value of $\frac{D}{SP^2}$ is evidently the same, whatever is the value of SP; & therefore its fluxion is the same that it would be if SP was constant; & therefore $= \frac{BS^2 \times B\dot{S}P}{2SP^2}$; & therefore $\dot{D}$, or the fluxion of $\frac{D}{SP^2} \times SP^2, = \frac{BS^2 \times B\dot{S}P}{2} + \frac{2D \times \dot{SP}}{SP}$

Therefore $\dot{A}$, or $\frac{3P\dot{D}}{\sqrt{SP}} - \frac{3PD \times \dot{SP}}{2SP^{3/2}}, = \frac{3P \times BS^2 \times B\dot{S}P}{2\sqrt{SP}} + \frac{6PD \times \dot{SP}}{SP^{3/2}} - \frac{3PD \times \dot{SP}}{2SP^{3/2}} = \frac{3P \times BS^2 \times B\dot{S}P}{2\sqrt{SP}} + \frac{9PD \times [\dot{SP}]}{[2SP^{3/2}]} = \frac{3P \times BS^2 \times B\dot{S}P}{2\sqrt{SP}} + \frac{3A \times \dot{SP}}{2SP}$ which is [...]

[...] made it before

The easiest way ⟨I know⟩ of finding the fluxion of A is by finding the fluxion of $\frac{3PD}{SP^2} \times SP^{3/2}$ as I did in the letter I sent you about it[1]

Note

For Henry Cavendish see note to Letter 10. This isolated sheet presumably represents his further response to Hutton's comments on his paper about finding cometary orbits.

[1]That is, Letter 10.

Letter 12.
CHARLES HUTTON to ROBERT HARRISON, 4 August 1779

Printed in Melmore, 'Some Letters', 74–6.

WOOLWICH, 4 *Aug.* 1779.

DEAR SIR,

Two or 3 days since I sent to the Durham waggon a parcel for you. It contains 7 copies of a paper I have lately had printed in the Philos. Trans.[1] which I desire the favour of you to distribute to the same Gentlemen to whom you delivered the copies of my former papers.[2] This last paper is a computation to ascertain the mean density of the whole globe of the earth; a very important problem, and a work of immense labour. You will perceive what the Data & Conclusions are, which I believe are not far from the truth, tho' it be the 1st experiment of the kind. I have accompanied them with one copy of the paper containing the astronomical observations,[3] which is intended for your own use, & is a proper companion to my paper. I have no letter from our friend Williams since I last wrote to you, but I hear he is well at New York. Mrs. Poole lives at Dover, & I believe is very well. Mr. Thos. Williams (their brother) is lately come from India, in a very bad state of health; he also at present I believe is at Dover.[4] Harry's (my son) station is at Coxheath Camp.[5] My own health &c pretty well as usual, so have nothing to tell you farther concerning myself, unless I mention that the University of Edinburgh have just now done me the honour to grant me a Diploma for the Degree of Dr. of Laws!

I forgot to mention above, that I could wish you to give one of the Copies of the paper to Mr. Hartley of Yarm,[6] & to present my respects to him.

Mr. Muller was himself in an error, instead of Mr. Robins, concerning the subject of his censure of him.[7] And now for my friend Mr. Emerson.[8] In the first place I must say that his whole conduct concerning the 2 probs of Bernouilli quite amazes me, nor could I easily have believed that a man of his knowledge & good sense could have committed so many inconsistencies & improprieties. I agree with you that the problems are little other than Algebraic Juggles, but still their solutions are not false, nor can the two quantities in the probs have any other possible meanings than one each, if letters & symbols have any determinate signification at all notwithstanding what Mr. E. says about the duplicity of the meaning. I like Bernoulli as little perhaps as Mr. E but in endeavouring to expose the former I was sorry that the latter exposed himself, for the method he made use of to do it appears to me highly improper even tho' B. had been in an error concerning these solutions. I repeat it therefore, it was with reason that I thought his conduct *extraordinary* & that it *surprised* me, for I hold it utterly unworthy the character he has long borne for mathematical knowledge. It has also given me some concern to find Mr. E. receive with evident ill-humour the well-intended hints of his friend. My time is too much employed for me to concern myself with any person whom I regard not, & I wished to give him an opportunity of correcting his oversights in a public manner himself & prevent his enemies from doing it for him in a disagreeable manner. And it is not impossible some such thing may happen, for since I wrote to you, I have received from a Diary Correspondent[9] a request to propose these very probs in that work on purpose to correct Mr. E's oversight; so that possibly such a thing may be sent to some other place where it may [be] used against him. But as I do not desire to trouble his repose, I wish not to communicate to him anything more on this subject, what I have farther to say on it being for your own satisfaction & at your own request. I shall just remark however that I had another motive for mentioning these probs to him, which was that as he possibly & unfortunately might not have a great number of years yet to live, & that as I foresaw I might have some care of any future editions of his books, I was desirous of procuring from himself any corrections or hints of improvements which he might wish to have made; which was a motive that had his honour for its object. And indeed I have already by his own consent the care of a new edition of his Algebra which is in the press.[10]

Let us first take an example in numbers to each of the probs viz., in the 1st prob. take $a = 6$, & let us approach to the sum of the series by beginning at the farther end of it & tracing it backwards. Now the last term of the series is $\sqrt{a}$ or $\sqrt{6}$ which is less than 3, i.e. $\sqrt{a} < 3$; take in the last term but one, then $a + \sqrt{a} < 6 + 3 < 9$, or $\sqrt{a + \sqrt{a}} < 3$; take in the antepenult. then $a + \sqrt{a + \sqrt{a}} < 6 + 3 \leq 9$, & $\sqrt{a + \sqrt{a + \sqrt{a}}} < 3$ and thus by continually adding another term the sum will be always less than 9, & $\therefore$[a] the root of the sum less than 3; so that the value of the series approaches continually to the No 3, which is its limit or the sum of the inf. series. So that in this instance the series is not only finite but = to B's theorem,

for $x = \frac{1}{2} + \sqrt{a + \frac{1}{4}}$ is $= \frac{1}{2} + \sqrt{6\frac{1}{4}} = \frac{1}{2} + \sqrt{\frac{25}{4}} = \frac{1}{2} + \frac{5}{2} = 3$.

Again take $a = 1$, & $b = 8$ in the 2nd prob. & trace the series back as before.[b] The[n] $\sqrt{b} = \sqrt{8}$ is < than 3, & $a \times \sqrt{b} = 1 \times \sqrt{8}$ less than 3, conseq. $\sqrt{a[\times]\sqrt{b}} < 2$, i.e. the last couplet is less than 2; repeat this operation, then

$b \times \sqrt{a \times \sqrt{b}} < 16$, & $\sqrt{b \times \sqrt{a \times \sqrt{b}}} < 4$ & $\sqrt{a[\times]\sqrt{b \times \sqrt{a[\times]\sqrt{b}}}} < 2$ i.e. the two last couplets less than 2; & the same thing happens after ever so many couplets; $\therefore$ the series approximates infinitely to the no 2, which is the value of B's theorem

$$x = \sqrt[3]{aab} = \sqrt[3]{1 \times 8} = 2.$$

Proceed we now to the investigations of the general theorems. And first for the 1st prob. Let $s^2 - s = a$, then shall s be the sum of the inf. series $\sqrt{a + \sqrt{a + \sqrt{a + \sqrt{a}}}}$ &c. For since $s^2 - s = a$, or $s^2 = a + s$, or $s = \sqrt{a + s}$, $\therefore s >$ (greater than) $\sqrt{a}$; add a to each, so shall $s + a > a + \sqrt{a}$, but $s^2 = s + a$, $\therefore s^2 > a + \sqrt{a}$, or $s > \sqrt{a + \sqrt{a}}$; add a again, then

$$s + a \text{ or } s^2 > a + \sqrt{a + \sqrt{a}} \ \& \ s > \sqrt{a + \sqrt{a + \sqrt{a}}};$$

then adding & extracting again, it will be $s > \sqrt{a + \sqrt{a + \sqrt{a + \sqrt{a}}}}$. And thus the series approaches continually to the value of s which is its limit. Since then $s^2 - s = a$, $\therefore s = \frac{1}{2} + \sqrt{a + \frac{1}{4}}$ = the sum of the inf. series.

Again in the 2nd prob. let us begin the operation at the farther end of the series, & trace its value backwards towards the beginning by actual multiplications & extractions. The last term of the series being $\sqrt{b}$ or $b^{\frac{1}{2}}$,

[a] Melmore consistently prints $\because$ where $\therefore$ seems to be intended.

[b] Melmore several times prints + where × was clearly intended.

let it be multiplied by a & the root taken, so shall $a^{\frac{1}{2}}b^{\frac{1}{4}} = \sqrt{a[\times]\sqrt{b}}$; mult. by b & take the root, then by a & take the root, so shall $a^{\frac{5}{8}}b^{\frac{5}{16}} = \sqrt{a\sqrt{b\sqrt{a\sqrt{b}}}}$; repeat the operation so shall $a^{\frac{21}{32}}b^{\frac{21}{64}} = \sqrt{a\sqrt{b\sqrt{a\sqrt{b\sqrt{a\sqrt{b}}}}}}$; in like manner the next operation will give $a^{\frac{85}{128}}b^{\frac{85}{256}}$; the next $a^{\frac{341}{512}}b^{\frac{341}{1024}}$; &c. where it is evident that the exponent of a is always double of that of b; that the exponent of a after any one extraction is always found by adding $\frac{1}{2}$ to $\frac{1}{4}$th of its exponent in the preceding expression; & that the exponent of b is found by continually adding $\frac{1}{4}$ to $\frac{1}{4}$th of its former one. But in all the expressions the exponent of b is a vulgar fraction whose denominator is always 1 more than three times its numerator, so that when its numerator is n its denominator is $\overline{3n+1}$,[a] & this property will continue without end, for $\frac{n}{3n+1}$ being any exponent of b, by the construction of the series the next exponent will be $\frac{1}{4} + \frac{1}{4}$ of

$$\frac{n}{3n+1} = \frac{1}{4} + \frac{n}{12n+4} = \frac{4n+1}{12n+4}$$

where the denom. is still $= 1+3$ times the numerator. Now when the series is supposed to be infinitely continued, n is infinite, & then the exponent $\frac{n}{3n+1}$ is $= \frac{n}{3n} = \frac{1}{3}$ the expon. of b; & $\therefore \frac{2}{3}$ the expon. of a; conseq. the value of the inf. series is $a^{\frac{2}{3}}b^{\frac{1}{3}} = \sqrt[3]{aab}$, as in the solution. Hence it appears that the sum of the series are finite & rightly assigned by Bernoulli.

I am, my Dr Sr with much respect & gratitude

Your much obliged & obedt Servt

CHA. HUTTON.

Notes

For Robert Harrison see notes to Letter 9.

[1]Hutton, 'An Account of the Calculations'.

[2]See the list of recipients in Letter 9.

[3]Presumably Maskelyne, 'An Account'.

[4]For Edward Williams see notes to Letter 9.

[5]Henry Hutton (?1761–1827), Charles Hutton's eldest child and only son: he graduated from the RMA in 1777 and had a distinguished career in the Royal Artillery; in retirement he became a keen antiquarian but, as a nineteenth-century historian would remark somewhat sourly, he 'lacked the constructive faculty' and his work never issued in publication: obituary of Henry Hutton in *Gentleman's Magazine* (December 1827), 561–2;

[a]Melmore mistakenly prints $3\overline{n+1}$.

Rogers, 'Register', 329; ODNB *s.v.* Charles Hutton; and see Wardhaugh, *Gunpowder and Geometry*, *passim*.

[6]Mr. Hartley of Yarm has not been identified.

[7]John Muller (1699–1784) was professor of fortification and artillery at the Royal Military Academy until 1766, and the author of several works on mathematical subjects: ODNB; MPHE. Benjamin Robins (1707–51) was a military engineer and an important writer on ballistics; Hutton's aim in his experimental work on the subject was to complete and extend that of Robins, and in 1805 he was responsible for a new edition of Robins' *New Principles of Gunnery* (see Letter 92): ODNB; see also Johnson, *Collected Works*. I have not located an occasion on which either man cites Hartley in print, and the 'censure' to which Hutton refers may have been a private one.

[8]For William Emerson, and his citation of problems by Bernoulli, see the notes to Letter 9.

[9]That is, a contributor to the periodical *The Ladies' Diary*, of which Hutton was editor from 1774 to 1817.

[10]Emerson, *Algebra*; there is no indication in the book that Hutton had a role in seeing this second edition through the press.

Letter 13.
CHARLES HUTTON to NEVIL MASKELYNE, 21 September 1779

Printed in Hutton, 'Calculations', 1–2.

The paper was read to the Royal Society on 11 November 1779: London, Royal Society, JBO/29, p. 489.

Royal Military Acad.
Sept. 21, 1779.

DEAR SIR,

AS the experiment of determining the universal attraction of matter, which you lately conducted with so much accuracy and success, is of so great importance and curiosity, that it may probably be hereafter repeated by the learned in other countries, and in other situations; and as the utmost precision is desirable in so nice an experiment; I have composed the inclosed paper to determine the best part of a hill for making the observations, so as to obtain the greatest quantity of attraction.

I have no doubt, Sir, that the determination of this point will appear of some consequence in the opinion of one who has the improvement of useful knowledge so much at heart; and if the manner in which it is here made, meet your approbation, I shall desire the favour of your presenting it to the Royal Society.

I have the honour to be, &c.

Notes

Nevil Maskelyne (1732–1811), astronomer and mathematician, was Astronomer Royal from 1765 until his death. He was one of the examiners who appointed Hutton to his position at Woolwich in 1774, and subsequently gave Hutton work on the *Nautical Almanac* as well as on the calculations related to his experiment at Schiehallion to determine the density of the earth, for which Maskelyne received the Copley Medal in 1774. ODNB; DSB; MPHE; Howse, *Nevil Maskelyne*; Higgitt, *Maskelyne*; also Letters 9, 17, 31, 76, 62, and 75.

Letter 14.
CHARLES HUTTON to ROBERT HARRISON, 31 May 1780

Printed in Melmore, 'Some Letters', 77–8.

WOOLWICH, 31 *May* 1780.

DEAR SIR,

I sit down to ask how do you do, without having anything material to tell you on my part, things here being in statu quo, & I have not had a letter from our friend Williams this 12 months. I sometimes however hear of him, & know that he is well, tho' he has had a bad & long fever since my last receipt.[1]

How does the old Gentleman at Hurworth?[2] Some months since I sent to him 7 Copies of an additional paper of mine on *the place of greatest attraction* on the side of a hill.[3] I sent them by a Mr. Wright,[4] a near relation of his, who was formerly a Coal Viewer about Tanfield &c. but has been for some years past in Scotland, where he was returning from London after being up on some business. The parcel was accompanied with a letter to Mr. E. in which, after keeping one of the pamphlets himself, I desired him to send the rest to you to dispose of as the former ones had been.[5] Have you received them, or heard of them? I often meet with Newcastle people in London, but the sight of my much esteemed friend H. would afford a real pleasure indeed. I met young Mr. Smith[6] lately: he is just going out to India. I am sorry to learn by him that the Grecian Lady[7] is in danger of losing that language again, or at least does not pursue it, her mother having assumed the whole direction of her, for fear her learning should spoil her Marriage! We have just come out the 2nd Vol. of Horsley's Newton (the Cuts all done by Beilby of N.C.).[8] Lately publish'd also a most excellent work on Infinite Series, by a Mr. Henry Clarke (not my friend Sam.).[9] Mr. Landen has just finished from the press a 1st Vol. of Mathematical Memoirs.[10] He intends going on if health permit, which is in but a delicate & doubtful way. His death would be indeed a great public loss, as he is a most excellent Mathematician & very worthy Man.

Old Peat, the author of the Gent. Diary,[11] is dead; & I have procured that work for the rev. Mr. Wildbore,[12] a near neighbour of a revd Mr. Geo. Walker[13] now at Nottingham, whom you may have formerly seen at Mr. Lowthian's or about N.C. after which he lived some time at Durham (being the Dissenting Minister) whence he sent some excellent probs & solutions to the Ladies' Diary under the signature of *P.M. of Durham.* I sometime hear the highest character of a Dr. Scott of Oxford, who I think was an old pupil of yours in N.C.[14] He is at present a Candidate (on the Whig Interest) for Member for Oxford University, as is also (the persian) Mr. Jones, author of the Synopsis Palmar. Mathes. & editor of Sir I. N's Miscellaneous pieces.[15] There is a 3rd Candidate, but I know not who. It is said Dr. Scott is safe for one.

Hoping you enjoy a good State of health, I remain

Your most sincere & affectionat hble Sert

CHARLES HUTTON.

Notes

For Robert Harrison see note to Letter 9.

[1]For Edward Williams see note to Letter 9.

[2]That is, William Emerson: see note to Letter 9.

[3]Hutton, 'Calculations'.

[4]Not identified.

[5]See the list of recipients in Letter 9.

[6]Not identified.

[7]This may be a punning reference to Hutton's daughter Eleanor ('Ellen'), her mother being Hutton's estranged wife Isabella; see note to Letter 16.

[8]Newton, *Opera*, in which the illustrations were made by the engraver and businessman Ralph Beilby (1743–1817), based in Newcastle for most of his career and better known as the master of Thomas Bewick (1753–1828). Bewick also had a connection with Hutton, having produced the engravings for Hutton, *Guide* in 1764. ODNB; 'Public Characters', 107; 'Memoir of Mr. Thomas Bewick', *Gentleman's Magazine* (January 1829), 17–20 at 18; Horsley, *Eighteenth-Century Newcastle*, 46.

[9]Lorgna, *Dissertation*, which was in fact dedicated to Hutton. 'My friend Sam' was one Samuel Clarke: during the period when Hutton's collection of problems from the

Ladies' Diary—the *Diarian Miscellany*—was publishing in numbers, he set up a rival publication under the title the *Diarian Repository*; it began in 1771 and included just the mathematical parts of the *Diary*, leaving out the other puzzles. After an undignified flurry of adverts and counter-adverts in the national newspapers, Clarke's *Repository* ceased publication in 1774, having reached only 1760 in its coverage of the *Diary*'s contents. See Archibald, 'English Mathematical Serials', 382 and Melmore, 'Some Letters', 77.

Henry Clarke (1743–1818), mathematician and schoolmaster, of Manchester, was one of those blocked from Fellowship of the Royal Society in the years preceding the 'Dissensions' of 1783–84, while in 1802 he was appointed to a teaching position at the Royal Military College at Marlow, on which appointment it is likely Hutton was consulted: ODNB, MPHE.

[10]Landen, *Mathematical Memoirs*; John Landen (1719–90) was a surveyor, land agent, and F.R.S., and wrote widely on mathematics: ODNB, MPHE.

[11]The *Gentleman's Diary* was, with the *Ladies' Diary*, one of the most important British mathematical periodicals of its time; it ran from 1741 until it merged with the *Ladies' Diary* in 1841: see Archibald, 'English Mathematical Serials', 382–3. Its co-founder and its sole editor from 1757 until his death was Thomas Peat (1707/8–80), also known as a teacher, lecturer, and surveyor: ODNB, MPHE.

[12]Charles Wildbore (1737–1802), mathematics teacher: MPHE.

[13]George Walker (1750–1807) was a distinguished mathematician and a friend of members of the Lunar Society; he was associated with the dissenting academies at Manchester and Warrington: MPHE. For Lowthian see note to Letter 9

[14]Melmore, 'Some Letters', 78 identifies Dr Scott as the judge and politician William Scott, later Lord Stowell (1745–1836), who unsuccessfully contested the Oxford seat in 1780: ODNB. Both Hutton and Harrison taught his younger brother John, later Lord Eldon: see Letter 133.

[15]William Jones (*c.* 1675–1749), author of *Synopsis palmariorum matheseos, or, A New Introduction to the Mathematics* (London, 1706) and editor of works by Newton (ODNB), was the father of Sir William Jones (1746–94), orientalist and judge: the latter campaigned for the Oxford seat in 1780 but withdrew before the election. Hutton confused the two men. The confusion is the more surprising as in his 1785 *Tables* (p. 118), Hutton clearly distinguished them.

Letter 15. CHARLES HUTTON to ROBERT HARRISON, 19 March 1781

Printed in Melmore, 'Some Letters', 78–9.

WOOLWICH, 19 *March* 1781

DEAR SIR,

I take the opportunity of my son's[1] going thro' Durham to send the papers accompanying this. He is going down to Scotland where he is ordered on duty, & has directions to wait upon you if the Stage Coach stop long enough in Durham to permit him so to do. I don't find that I have much new intelligence to give you. Our friend W.[2] was very well at New York pr. last Accots & in a way of making money, so the longer the conquest of America is protracted from year to year, the better for him & most others concerned in it. How is Mr. Emerson?[3] I know not of any thing in the Transactions about the rotation &c of Venus, but there was lately read a paper at the Society, from a Mr. Herschel at Bath,[4] concerning his observations of the rotation &c of the planets, which paper I believe will be printed in the Transacs. The Society have begun their meetings at their new apartments at Somerset House, which are more spacious & elegant than the former. The account of the last voyage on Discoveries is already drawn up by Cap. King, one of the 2 Commanders who brot the two ships home after the death of Cap. Cook,[5] but probably we shall not see it published these 2 years, owing to the delay of the plates, which are to be executing in the very best manner by the best engravers. You will perhaps smile when I tell you I am become *Astrologer General* to the Stationers' Co having entered into agreemt with them for the sole conducting of all the Almanacs. Not that I can execute them all myself, but shall likely employ assistants to make most of the calculations & certainly all the Prognostications, these belonging to a Science far above my pitch.[6] The conduct of the almanacs had been in the family of the Wings, from father to son, ever since the famous Vincent Wing Astronomer & Astrologer,[7] & has only ceased with the death of the late Tycho Wing the last year; the present representative of the family being a young man not at all concerned in such matters, the Co [thought] that it would be proper to make a change. Speaking of these things brings to my mind a good anecdote of a young attorney of N.C. whom we know very well which I shall here copy in the words of my correspondent who lives near the Mr. Wright in Nottinghamshire mentioned below, which Mr. Wright has for many years written the Prognostications in Moore's Almc[8] & has the character of a great conjurer, telling the people their fortunes &c. & to this Sidropher[9] it seems our curious townsman applied himself. My correspondent's words are, "Mr. Wright shew'd me a letter from a gentleman of Newcastle, I think a Mr. H—— an attorney, giving him an account of his second marriage & elegant situation. This gentn, some years since, sent to Mr. Wright to calculate his nativity; and he, like all other fortune-tellers, told him great things. Afterwards having paid his addresses to two ladies at once, he sent to Mr. Wright, who had told him he was to have two wives, to know whether he should marry A or B; he ordered him to marry A, which he did & in the letter which I saw he acquaints Mr. Wright with her death, & that he had just married B"! This may afford a laugh at N.C. sometime.

I remain,

Dr Sr Yours most sincerely,

CHA. HUTTON.

I send pr my son 2 copies of the paper on Cubic Equas[10] to Mr. Anderson of N.C.[11]

Notes

For Robert Harrison see note to Letter 9.

[1]Henry Hutton: see note to Letter 12.

[2]Edward Williams: see notes to Letter 9.

[3]For William Emerson see notes to Letter 9.

[4]For William Herschel see note to Letter 21. His paper, read to the Royal Society on 11 January 1781, was printed as Herschel, 'On the Rotation of the Planets'.

[5]James King (1750–84) accompanied the third voyage of the well-known explorer James Cook (1728–79) as second lieutenant and astronomer. After the death of Cook and another commander he took command of the *Discovery*; he assisted in preparing Cook's journal of the voyage for the press (Cook and King, *Voyage*), and his narrative of the final part of the voyage formed the third volume: ODNB.

[6]The Stationers' Company of London held a monopoly on the printing of almanacs at this time. Documents in its archive show that the agreement mentioned here was one of several: in 1775 Hutton committed himself to the compilation of certain almanacs (we do not know which, but the *Ladies' Diary* was certainly included), and in 1786–7 his agreement was renewed again. By then it brought him 130 guineas per year but excluded the *Gentleman's Diary*, which presumably remained in the hands of Charles Wildbore (see Letter 14); it also excluded *Poor Robin's Almanack* and possibly one or two of the single-sheet almanacs. In 1788 an agreement was drawn up in which Henry Andrews (1744–1820) undertook the actual compilation of an unspecified list of almanacs for a paltry £20, Hutton remaining as their better-paid overseer. Myers, 'The Records of the Stationers' Company', Court Book M, p. 483; Series I, Box C, folder 4; folder 6, items i–iv; Court Book O, p. 77; also ODNB *s.v.* Andrews, Henry; and a notice of Andrews' death in the *Monthly Magazine* (December 1820), 484–5.

[7]Vincent Wing (1619–68) published major works on astronomy but is best remembered as the founder of the dynasty of almanac-makers alluded to here, which included his sons Vincent and John (1662–1726), his grandson Tycho (1696–1750) and great-grandson Tycho (1726–80): ODNB.

[8]That is, the long-running work of semi-serious astrological prognostication known variously as *Vox Stellarum*, *Moore's Almanac*, *Old Moore*, or *Old Moore's Almanac.*

[9]Properly Sidrophel, alluding to the comic astrologer in Samuel Butler's satire *Hudibras* (1663).

[10]Hutton, 'Of Cubic Equations'.

[11]Possibly George Anderson (*c.* 1726–98), for whom see note to Letter 9.

Letter 16. EDME-SÉBASTIEN JEAURAT to the MISSES HUTTON, 20 October 1781

Washington, D.C., Smithsonian Institution, MS 752 A: ALS.

A l'observatoire. A Paris ce 20, octobre 1781.

Mes demoiselles

Je ne saurois vous exprimer jusqu'à quel point j'ai été sensible à l'honneur de vôtre souvenir, je vous demande à toutes deux ainsi qu'au cher Papa la continuation de ce bon souvenir de moi et quant j'yrai en Angleterre comme je l'espere dans 4 ans je profiterai ⟨de⟩ la permission que vous me donnez, d'aller vous voir; à lors j'aurai l'honneur de cultiver plus amplement l'honneur de votre connoissance.

Ce que je vois avec plaisir Mes demoiselles, c'est que vous écrivez le françois aussi bien que vous le parle; [*sic*] je desirerois qu'il en fut ainsi de moi à legard de l'Anglois mais je suis trop vieux pour faire cette entreprise et je mourerai ignorant comme je l'aurai eté toute ma vie.[1]

En me mettant à vôtre place Mes demoiselles je sens tout l'agrement que que [*sic*] vous avez, avec un aussi bon Papa que le vôtre; actuellement il vous promene[a] dans votre pays même, et ce pay que j'ai trouvé si beau, doit etre presque pour vous un pay nouveau car vous en êtes sortie [*sic*] très jeunes;[2] Embrassez s'il vous plais Mesdemoiselles toutes deux vôtre très-cher Papa pour moi, et quant j'yrai à Woolwich je vous rendrez à chacune ses Embrassades; d'avance je m'en fais une feste et c'est avec reconnoissance et respect que j'ai l'honneur d'etre,

Mes demoiselles,

Vôtre très-humble et très-obeissant serviteur

Jeaurat

A Mesdemoiselles
Mesdemoiselles Hutton
chez leur respectable et charmant Papa
A Woolwich près Greewich.

---

[a] *altered from*: promenez.

Notes

Edme-Sébastien Jeaurat (1724–1803) was a Parisian astronomer based from 1770 at the Royal Observatory: DSB. See also Letter 21, which indicates that Jeaurat's correspondence with Hutton continued at least until 1784.

Jeaurat addressed two (young) women, whom he had evidently met on a visit to Woolwich and from whom he had received a letter subsequently: presumably Hutton's two oldest daughters Isabella and Camilla, aged 19 and perhaps 16 at this date. Isabella (1762–1839), Hutton's second child, never married and was by 1819 Hutton's 'constant amanuensis' and companion following the death of his second wife; her role in his work at an earlier stage is unknown. Camilla (?1764–94), Hutton's third child, married Charles Henry Vignoles in 1790 (see Letter 49) and died with him of yellow fever in Guadeloupe in 1794, having borne in 1793 a son, Charles Blacker Vignoles (for whom see note to Letter 61).

This letter is the earliest piece of evidence placing either woman in Woolwich, where they had moved ('très jeunes', according to Jeaurat) from the home near Newcastle of their mother Isabella Hutton (1733/4–85; she was married to Charles Hutton in April 1760 and died on 26 May 1785, never having moved away from the north-east). The absence of any reference to Hutton's third daughter Eleanor, aged 12, may imply that she was still in the north-east. Eleanor (1769–?1850) was married three times, to an O'Leary, to Captain Henry Wills, and to George Children, but she makes few appearances in the surviving correspondence and her whereabouts at various periods is obscure; it is not certain whether she had any children.

Charlotte, the daughter of Hutton's second wife Margaret Ord (see note to Letter 67), was certainly a member of the household at Woolwich at this time, but at three was presumably too young to receive Jeaurat's notice.

[1]This passage implies indirectly that Hutton himself spoke some French (implied also in a letter of Margaret Hutton to Charles Blacker Vignoles of 13 December 1816—Portsmouth, History Centre, Vignoles Papers, Letter 126—where Margaret lamented the absence of Hutton and Isabella when a French astronomer visited and implied that unlike her, they would have been able to speak his language); for his study of languages see Letter 9. That he was keen to expose his daughters to French language and culture is emphasized by the fact that the two youngest, Eleanor and Charlotte, subsequently lived for a period in France: Obituary of Charlotte Hutton in *The Gentleman's Magazine* (October 1794), 960–1; letter of Eleanor Wills (née Hutton) to Charles Blacker Vignoles, December 1816, in Portsmouth, History Centre, Vignoles Papers, Letter 127.

[2]It is stated in Bruce, *Memoir*, 22, that Hutton did not in fact revisit the north-east at this or at any other time.

Letter 17.
LEWIS EVANS to CHARLES HUTTON, 12 February 1782

Oxford, Museum of the History of Science, MS Evans 31: draft.

Lewis Evans' letters to Hutton survive in the form of a bound volume of Evans' drafts, which also contains a few other items. Many of the letters are covering notes for proposed problems or solutions for the *Ladies Diary*; in such cases the problems and solutions themselves are not included in this edition.

Compton near Torringdon Berks
12th. Feb. 1782.

Sir.

I have attempted to answer the two first questions of the Ladies Diary ⟨time wd. not permit me to attend to more⟩. If the subjoined solutions should be right, you have my leave to make what use you please of them: & I shall be happy in the Idea of having in so small a measure, contributed towards the utility of that vehicle of mathematical knowledge

I am &c

L. Evans

...

For the Ladies Diary
stationers hall
London

Notes

Lewis Evans (1755–1827) was a clergyman and teacher in, successively, Compton, Hungerford, and Little Bedwin, villages within a few miles of each other in Berkshire and Wiltshire; he evidently taught mathematics, but in Letter 42 it is implied that he also taught 'the Classics', at least in Hungerford. Initially he wrote to Hutton as the anonymous editor of the *Ladies' Diary*, but after he received a personal response during 1783 (see Letter 23), theirs became a more personal exchange marked by what Evans called (Letter 41) 'many Civilities and kindnesses' from Hutton and hares and pheasant from Evans. It continued until (at least) 1794; in 1799 Evans became a mathematical master at the Royal Military Academy under Hutton. Later in life he would become known in a modest way as an astronomer, and he was elected F.R.S. a few months after Hutton's death in 1823. ODNB; MPHE; also letters 23, 30, 33, 34, 40, 41, 42, 43, 44, 45, 46, 47, 48, 50, 51, 53, 54, 56, 57, 58, 63, 64, and 65.

Letter 18. CHARLES HUTTON to NEVIL MASKELYNE, 12 August 1782

Printed in Hutton, 'Project for a New Division of the Quadrant', 21.

The paper which this letter accompanied was read to the Royal Society on 27 November 1783: London, Royal Society, JBO/31, p. 260.

Royal Mil. Acad. Woolwich,
Aug. 12, 1782.

DEAR SIR,

HAVING long since thought it would be a meritorious and useful service, to adapt the tables of sines, tangents, and secants, to equal parts of the radius instead of to those of the quadrant; and having frequently mentioned this project to you, SIR, as a proper judge and promoter of all useful improvements in science; I now beg leave to lay before you some observations I have thrown together on the subject, with a view to stimulate others, either to undertake and calculate some part of so large and painful a work, or to communicate farther hints for the improvement and easier performance of it.[1]

I have the honour to be, &c.

Note

For Nevil Maskelyne see note to Letter 13.

[1]Hutton's intention to organize a team of calculating assistants to construct trigonometric tables in radians rather than degrees was mentioned subsequently in the introduction to his 1785 *Tables*, and (perhaps by mistake) also in the introduction to the 1794 second edition of that work. The proposed new tables themselves never appeared, however; see also Letter 46.

Letter 19.
JOHN PLAYFAIR to CHARLES HUTTON,
12 December 1782

London, Senate House Library, MS 913B/3/1 (xiv): ALS.
Endorsed 'Dr Olinthus Gregory's Sale March 18, 1842'.
Sender reported in further endorsement.
The enclosure mentioned is not extant.

Edin^{r} 12th Decr. 1782

Dear Sir

You will form a very indifferent opinion of my Industry, when I tell you that every week since I saw you at Woolwich I have intended writing you, and inclosing the Theorem I mentioned at that time, without ever being able to execute that intention. A considerable change that has taken place in my way of Life and my removal from my house in the Country to this place[1] have occupied me so entirely, that I have studied less for these five Months past, than at any other period for several years. The Propositions I send you are not after all of any great consequence. The first is to be met with in a Paper by Euler somewhere in the Petersburg Commentaries, but embarassed by an Algebraic Calculus most injud[ic]iously instituted; and I give it you as an instance in which Geometry is much safer than Algebra. Not that the conclusion might not be brought out by the last of these methods as well and as simply as by the former, if proper attention were bestowed; but that considering the incautious way by which Analysts usually proceed, there is a great probability that the Geometer will succeed the best of the two.

I have seen your Son[2] only for a moment, as he left Edin^r to return to the head quarters at Perth, much about the time that I came to it. He was very well however and in good spirits when I saw him.

We hear wonderful things of Mr Herschel's observations on the Moon, and it is said that he has discovered symptoms[a] of <u>Animation</u> in that Planet.[3] What may be understood by this I cannot tell, but all of us here are very impatient to be more fully informed.

Engaged as you are in a variety of Literary pursuits, you can have little leisure for correspondence, and I therefore dare not venture to sollicite it. It would give me however the greatest pleasure to hear from you sometimes and to be informed how you are, and what Mathematical speculations are going forward with yourself or your freinds. In return you may expect, not indeed much literary intelligence, but much thankfulness, and now and then if you desire it, a Geometrical or Algebraical investigation.

If you insert the Prop^s. I have sent you in the Diary, you may affix to them any name you please that will conceal my own, that of <u>Atticus</u> for instance.[4] I had forgot to tell you that the Locus, so far as I know is new, except that I proposed some years ago to Mr Glenie,[5] and any value it may have is not from its difficulty, but from the great number other propositions that may be derived from it.

You will probably have D^r <u>Trail</u>,[6] with whom I think you are acquainted, in London, this winter, and I am very much persuaded that you and many other of my Freinds will have great Pleasure in his Conversation.

Prof. Stewart[7] begs to be remembered to you. Be so good as present my compliments to Miss Hutton & her sister.[8] I am much obliged to you for the

[a] *altered from:* An.

Civilities you have shewn my Brothers[9] since I left London, and I remain, sincerely yours

[*signature has been cut off*]

P.S. If you are so go [...][a] N° 25 St. Andrews Square Edin[r].

Doctor Hutton
Royal Academy Woolwich

Notes

John Playfair (1748–1819), mathematician and geologist, visited London in 1782 and was introduced to various prominent natural philosophers and mathematicians; it was probably during this period that he became acquainted with Hutton. He had also visited Nevil Maskelyne at Schiehallion during 1774–5, and would later support Hutton by undertaking, at his instigation, a lithological survey of the Schiehallion site in 1801. From 1804, however, as a reviewer and author of articles in periodicals and encyclopedias, he would become outspoken as a critic of the state of British mathematical science which Hutton to some degree represented. ODNB; DSB; MPHE; Playfair, *Works*, vol. 1, pp. lxxvii–lxxxviii; see also Letters 39, 60, and 120.

[1]Playfair was by now living in Edinburgh; see later and in the postscript to this letter.

[2]For Henry Hutton see note to Letter 12.

[3]For William Herschel see note to Letter 21; as Michael Hoskin notes in ODNB, the draft of Herschel, 'Astronomical Observations Relating to the Mountains of the Moon' 'included highly inappropriate speculations about lunar inhabitants'.

[4]The 1784 *Ladies' Diary* (p. 47) contained a single question by 'Atticus'; the following year (p. 38) his own solution to the problem was printed, but 'Atticus' made no further appearances in the *Diary*: index in Leybourn, *Mathematical questions*, vol. 4. The question required a proof of the following:

> In every triangle, the intersection of the perpendiculars let fall from the angular points on the opposite sides, the center of gravity of the triangle, and the center of the circumscribed circle, are all in the same straight line; and the distances of those points from one another, and in a given ratio.

[5]Probably James Glenie (1750–1817), soldier and mathematician, who had studied at the University of St Andrews and would vigorously support Hutton in the 'Dissensions' at the Royal Society in 1784: ODNB.

[6]Probably William Trail (1746–1831), who had formerly held the chair of mathematics at Marischal College, Aberdeen, but was at this time working in Connor as a clergyman in the Church of Ireland. He was the author of an algebra textbook and, later, a life of Robert Simson: ODNB.

[7]Perhaps Dugald Stewart (1753–1828), who at this time held the chair in mathematics at Edinburgh University; in 1785 he would move to the chair of moral philosophy, the field

[a]text lost due to cutting of paper.

with which he would come to be most associated: ODNB. In O.J. Vignoles, *Life of Charles Blacker Vignoles*, 45 he is reported to have known Charles Blacker Vignoles as a child, implying that he visited Hutton in Woolwich or London between about 1795 and 1815.

[8]Playfair presumably referred to the two oldest of Hutton's four daughters; see note to Letter 16.

[9]Playfair had four brothers, including James (1755–94), architect, and William (1759–1823), the writer on political economy who invented statistical graphs: ODNB. The former may have been and the latter certainly was in the London area at this time.

Letter 20.
CHARLES HUTTON to an unknown correspondent, 19 May 1783

Durham County Record Office, D/Gr 461 [M 11]: ALS.

Cover lost.

Woolwich, 19 May 1783.

I am very sorry, my Dear Sir, on my own accot. that the Knocker had become invisible again. I am not much in town, tho' still more than I wish. I am glad however you are so near me, as I was at a loss to think what was become of you this so long time. Am pleased also at the intended visit you promise to favour me with. And wish much you will come here to meet some of our Math[l]. & Philosophical friends at this place on Friday this week.[1] Our friend Williams[2] is expected here now very soon. I know not if you now amuse yourself sometimes with taking a peep at our friends above, if you do, there is an uncomon affair attending the star Algol that has lately been discovered by a young man of the name of Goodrick in York, who is deaf, & was dumb, but who has been taught to speak, read & write by Braidwood of Edinburgh.[3] The accident is a sort of eclipse which that star suffers periodically every 2 days 21 hrs, & which eclipse lasts about 7 hours each time, during $3\frac{1}{2}$ hrs it decreases in splendor, & $3\frac{1}{2}$ hrs it increases again. The star is naturally of the 2d magnitude, but decreases so as to appear of the 4th mag!

Adieu.

Yours affectionately

Cha. Hutton.

Notes

[1]After the 'Dissensions' at the Royal Society in 1784, Hutton and certain others were involved in a 'Friday Club' meeting at the Mitre Tavern on alternate Fridays; see Letters 28, 52, 62, and 87, and Howse, *Nevil Maskelyne*, 161. This passage seems to imply that Friday meetings were already taking place at Hutton's home in Woolwich by this date, although it is not clear whether attendance extended beyond the Woolwich–Greenwich circle.

[2]For Edward Williams see note to Letter 9.

[3]Thomas Braidwood (1715–1806), who had earlier run a mathematical school, is remembered as a teacher of the deaf and dumb in Edinburgh and later London: ODNB. John Goodricke would publish his observations on Algol as Goodricke, 'A Series of Observations' and Goodricke, 'On the Periods of the Changes of Light in the Star Algol', and there were subsequent papers, also in the *Philosophical Transactions*, by Johann Moritz de Brühl and Henry Englefield.

Letter 21. CHARLES HUTTON to WILLIAM HERSCHEL, 6 February 1784

London, Royal Astronomical Society, William Herschel Correspondence 13.H.35: ALS.

Woolwich
6 Feb. 1784.

Doctor Hutton presents respectful Compliments to Mr Herschel, hopes a continuance of his health & good spirits enable him to continue his valuable observations & labours in the improvement of Science, & sends him from Mr Jeaurat of Paris[1] some new tables of the Planet Herschel, as the book of which they are a part will not be published till the end of this year.

Notes

William Herschel, musician and astronomer (1738–1822), described his own life up to 1784 quite fully in Letter 26; he became a Fellow of the Royal Society in December 1781 and may well have first encountered Hutton there (he worked in Newcastle as a musician during Hutton's residence there, but Hutton's reference to 'a Mr. Herschel' in Letter 15 seems to indicate that they were not acquainted at that early stage). Their correspondence would subsequently become more cordial (Letters 25, 26, 27, 28, 29, and 55) and Hutton would describe him after his death as 'my very old and worthy friend' (Letter 132). The extraordinarily stiff tone of this note may have something to do with the 'Dissensions' that were unfolding at the Royal Society during exactly this period (November 1783–February 1784): it is not known whether Herschel supported Hutton, Banks, or neither.

[1]For Edme-Sébastien Jeaurat see note to Letter 16. He was at this time editor of the French astronomical ephemerides the *Connoissance des temps*; the volume for 1786 (printed in 1783) carried a brief notice of the discovery of 'la Planète Herschel' (pp. 3–4), while the volume for 1787 (printed in 1784) included tables for its position alongside those of the other planets in the monthly calendar (pp. 25, 29, 43, 57, 71, 85, 99, 113, 127, 141, 155, 169), as well several separate tables of its position and movements (pp. 176–204). The calculations had been performed by Nicolas-Antoine Nouet (1740–1811). (The following year the editorship passed into other hands and the longer separate tables for the new planet were not repeated, though brief predictions continued to appear.)

Letter 22.
JOSEPH BANKS to CHARLES HUTTON, ?April 1784

London, Wellcome Collection, MS 5270 no. 5: ALS.
Date from internal evidence.

Dear Sir

Col Roy Jnr De Luc &c.[1] & myself intend to pay Woolwich a visit Early on Wednesday morn in order Geometricaly & Barometricaly to measure Shuters hill.[2] you may suppose that an undertaking in which you can so materialy assist could never be undertaken without you[r] assistance being requested I have therefore undertaken in the name of Phipps Roy De Luc self &c to inform you that we intend favente cælo to be at woolwich by 9 AM on wednesday next when we shall immediately repair to your Quarters in order to request your advise assistance &c to measure a base somewhere below proper to determi[ne] the hight of the hill &c &c.

adei[u] Your affectionate
Servant JBanks.

Professor Hutton
Royal Academy
Woolwich

Notes

Sir Joseph Banks, baronet (1743–1820), naturalist and patron of science, was President of the Royal Society from 1778 until his death: ODNB, DSB, MPHE. Banks signed the certificate for Hutton's election to Fellowship of the Royal Society in 1774, but we do not know when or how they became acquainted. Relations between the two had reached an icy standoff by about 1780 and broke out in Hutton's forced resignation from his position of Foreign Secretary (on which position see Letter 9) and the ensuing 'Dissensions' at the

Royal Society in 1783–4. See Wardhaugh, 'Charles Hutton and the "Dissensions"' and the introduction to this volume. Letter 79 shows the two men corresponding in 1797, but in about 1815 Hutton referred to Banks as his 'implacable enemy' (Letter 115), while publications by Hutton's protegé Olinthus Gregory after Banks's death characterized him as the consistent enemy of the whole circle of (British, metropolitan, practical) mathematicians: [Gregory], 'A Review'.

Banks did not date this letter; a pencil endorsement of unknown authorship places it in 'early 1784'. In fact it can be dated with some confidence from other evidence. William Roy began to survey certain sites in and near London during 1783; by October he had been assigned responsibility for the English half of the international survey connecting Greenwich and Paris, and his systematic triangulation of southern England began on 16 April 1784 with a baseline on Hounslow Heath, near Greenwich, used Shooter's Hill as one of the subsequent bases of triangulation, and concluded at Dover late in 1787: see Jungnickel and McCormmach, *Cavendish*, 315. It is therefore just possible that the visit of Banks, Roy, Deluc, and Phipps to Shooter's Hill took place during 1783, but a date in or soon after April 1784 seems much the most likely.

[1]Banks referred to William Roy (1726–90), surveyor and founder of the Ordnance Survey (ODNB, MPHE); Jean-André Deluc (1727–1817), natural philosopher and politician (ODNB); and, below, Constantine John Phipps, second Baron Mulgrave in the peerage of Ireland and Baron Mulgrave in the peerage of Great Britain (1744–92), naval officer and politician and a friend of Banks (ODNB).

[2]Shooter's Hill stands a mile or so south of Woolwich; Hutton would live there for a period around 1786.

Letter 23.
LEWIS EVANS to CHARLES HUTTON, 22 April 1784

Oxford, Museum of the History of Science, MS Evans 31: draft.

Compton near Farringdon
Berks
April 22^{d}. 1784

Sir

[a]If the following solutions meet with your approbation they are at your service

[a] *deleted*: Should you think the following solutions to claim any merit I make no doubt but they will meet with their proper notice & direct.

I am, Sir, your very humb Ser[t].
L: Evans

...

I shall be much obliged by the insertion of the following questions in any future diary, provided you think them[a] worthy of a place in it.

...

That you might not be deceived with regard to the origin of this equation I think it proper to tell you that it is the final equation given by a M[r]. Mallock as a solution to question 478 of the Ladies' Diary 1762,[1] where you may see that he brings out the value of z to be $\frac{3a}{9}$. I have often tried to solve his equation but without any success:[b] it is for this reason I ventured to send it to you, in hopes you would propose it to public consideration. Indeed if it were not too great intrusion I should esteem it as a particular favour if you re-proposed the question itself in your next diary for a more satisfactory algebraic solution to be given it. I wish this partly on account of the above equation; and partly, as I see an elegant & concise solution might be exhibited from the conclusion you or your predecessor made under what M[r]. Mallock has shewn[c].

...

I cannot let this opportunity pass without returning ⟨you⟩ my[d] acknowledgements for the obliging Letter I received ⟨from⟩ you[e]
I am Sir
your very humble Servant
L: Evans
To the ingenious complier of[f] the Ladies Diary.
1783

---

[a] *altered from*: they are.
[b] Deletion obliterated.
[c] *altered from*: done.
[d] Deletion obliterated.
[e] *deleted*: and for the ?notice you?
[f] *altered from*: Gentleman who compiles.

Note

For Lewis Evans see note to Letter 17.

[1]The philomath periodicals such as the *Ladies' Diary* generally insisted that material submitted for publication should be original, and regularly named correspondents who were found to have offended in this respect; Charles Hutton, writing as 'Tonthu', had been thus reprimanded in *Martin's Magazine of Sciences* (August 1762), pp. 914–15.

Letter 24.
CHARLES HUTTON to KLEINSCHMIDT, 13 May 1784

London, Royal Society, RSL/2, no. 27, fol. 2v: draft.
This draft reply is written inside the cover of Kleinschmidt's letter to the Royal Society, dated from Arolsen in the Principiality of Waldeck on 21 March 1784.

Londres le 13. May

Monsieur

La Societé Royale de Londres vient de recevoir la Lettre que vous avez bien voulu lui adresser du 21 Mars dernier, y-joint la tres curieuse dissertation du Prof: Lohmeier de Artificio Navigandi per Aerem, dont il vous a plu lui communiquer non seulement une Copie de la reïmpression recente avec une traduction Allemande, mais meme votre exemplaire tres rare[a] de l'Edition de 1676.[1] Je suis chargé Monsieur de la part de la dite Societé de vous temoigner ses remercimens les plus particuliers pour ce present que les circonstances actuelles rendent effectivement des plus interressans. Je regrette au reste dans cette occasion que la reserve que cette Societé se prescrit comme Corps[2] me prive du plaisir de m'etendre ici sur la sensation que cette brochure a produit sur les individus qui[b] l'ont examinée — Je profitte cependant avec bien de l'empressement de la Conjoncture pour vous prier d'Agréer les Assurances[c] de la haute Consideration &c
CH

[a] *altered from*: presque unique.
[b] *altered from*: que.
[c] *altered from*: sentimes.

Notes

I have not succeeded in identifying Kleinschmidt, who described himself in his letter to the Royal Society of 21 March as 'Conseiller de Régence et du Consistoire de s. a. s. le Prince de Waldeck' and was described in the book he enclosed with that letter as 'Justizrath Kleinschmidt in Arolsen'; neither source gives his first name.

This appears to be the only extant example of the letters Hutton was supposed to write as Foreign Secretary of the Royal Society, and it is curious that it dates from after his resignation from that role the previous winter. Item 36 in the same volume of Royal Society correspondence bears the endorsement 'Answd May 5 1784 CH', probably indicating that Hutton had written a similar letter—not extant—to A.J. Renaux at Paris in reply to his of 6 January. This is also the only certain specimen of Hutton's written French (but see Letter 120, whose French version *may* be of Hutton's composition): it is always comprehensible if occasionally awkward, *dans cette occasion* and *ce present* being the most notable lapses.

[1]The work alluded to was the 1676 Wittenberg dissertation of Philipp Lohmeier (1648–80) 'De artificio navigandi per aërem', published at Rinteln in 1676. A new edition appeared at Arolsen in 1784 with a parallel German translation, for which Kleinschmidt was responsible.

[2]Hutton presumably alluded to the printed form of thanks with which the Royal Society acknowledged gifts, intending to avoid any appearance that judgement was being expressed or endorsement conferred; Hutton as Foreign Secretary completed and signed such forms, and it seems to be implied that Kleinschmidt's gift had been acknowledged in this plain style. This 'reserve' had become a difficult subject, however; in the context of the 'Dissensions' of 1784 one anonymous pamphleteer stated that it was clearly understood he should often add a few words of *politesse* on his own account, and evidence reached Joseph Banks that his failure to do so had given offence: see Wardhaugh, 'Charles Hutton and the "Dissensions"'.

Letter 25.
CHARLES HUTTON to WILLIAM HERSCHEL, 12 November 1784

London, Royal Astronomical Society, William Herschel Correspondence 13.H.36: ALS (our source).
Also printed in Gingerich, 'William Herschel's 1784 Autobiography', 74–5.

Woolwich, Nov. 12, 1784.

Dear Sir

I last night learned, by mere accident, from a gentleman concerned in the Composition of the European Magazine, that the proprietors of that work[1]

have caused to [be] engraved a portraite of you, which they intend[a] to insert in the very next N°. of that work, to be published the 1st of December, together with with [*sic*] such anecdotes[b] or account of your life as they can collect on this occasion.[2] This, Sir, you are to understand, is the ⟨usual⟩ practise[c] of that Magazine, not to satyrise or abuse Gentlemen, but to do them honour. Observing therefore your many eminent discoveries, they have taken this course with a view to do honour to your great merits, by exhibiting to the public the portraite & anecdotes[d] of so remarkable a personage. Now, Sir, as there is no avoiding this exhibition, even if you should ever so much desire it, and as ⟨they⟩ can pick up but very few particulars relative to your birth, & life, &c. and those they may meet with in public may probably be most of them errors, I would advise you, as the best way, to furnish them yourself with some anecdotes relative to accounts & dates of your birth, parents, country, studies, life, discoveries, &c. &c. which will be worked up, & put into order in a way to do you the most honour. I desired them to desist a few days till I should inform you, as I now do; and if you will send me any such account as above, I shall take care to see the draft made out from them before it is published, so as to set you in the fairest and most respectful light. You will please to observe that the utmost dispatch is necessary, as the thing must be printed this month, & published the 1st of Dec[r]. next. —
I am

Dear Sir,

Your sincere friend & hble. Servant.

Cha: Hutton.

William Herschel Esqr,
Datchet
near Windsor

Notes

For William Herschel see note to Letter 21. The matter of this letter is further addressed in Letters 26, 27, 28, and 29.

[1]At this date the *European Magazine, and London Review* was edited by Isaac Reed (1742–1807): ODNB.

[2]The planned account of Herschel eventually appeared in 1785 as Herschel, 'William Herschel'.

---

[a] *altered from*: intent.
[b] *altered from*: annecdotes.
[c] *altered from*: practice.
[d] *altered from*: annecdotes.

Letter 26.
WILLIAM HERSCHEL to CHARLES HUTTON, November 1784

Cambridge, MA, Harvard University, Houghton Library, Sir William Herschel Papers, MS Eng. 1414, f(29): copy in the hands of William and Caroline Herschel (our source) (see Gingerich, 'William Herschel's 1784 Autobiography', 77).
Printed with modifications in Herschel, 'William Herschel'; also printed in Gingerich, 'William Herschel's 1784 Autobiography', 77, 79, 80.

In a letter to D[r]. Hutton.

I was born in the City of Hannover Nov[r]. 15. 1738. My Father was a Musician & I was the second of four Sons whom he all brought up to the same business from their very infancy. My three Brothers are still in the same vocation, the eldest & youngest being engaged in his Majesty's Orchestra at Hannover & the third residing at Bath; all of them remarkable not only for their Genius in Music but much estimed as men of a valuable charactre. Having also two daughters, so numerous a family would not permit my Father, with his scanty circumstances, to bestow much on the education of his children & the utmost he could do was that besides the common learning of a school such as reading, writing & arithmetic, he provided me with a private instructor for the french language. The person who undertook this care finding me of a very ready memory so as to give him no trouble in that line, and being himself a man of Science chiefly devoted his time to encourage the taste he found in his pupil for the study of philosophy, especially Logic, Ethics & Metaphisics which were his own favourite pursuits, To[a] this fortunate circumstance it was undoubtedly owing that altho' I loved Music to an excess & made a considerable progress in it, I yet determined with a sort of enthusiasm to devote every moment I could spare from business to the pursuit of knoledge which I regarded as the sovereign Good, & in which I resolved to place all my future views of happiness in life. Thus several years were spent in study & in the practise of my profession, to which I was introduced at a very early period; till the troubles in the Electorate of Hannover, during the last war, made my situation there very uncomfortable.[1] The known encouragem[t]. given to Music in England determined me to try my fortune abroad & accordingly ab[t]. the year 1759[b] I came to settle in

[a] *altered from*: For.

[b] A pencil insertion corrects this to *arrived in Nov[r] 1757.*

this country where I had before passed some months in company with my Father & eldest brother being then in his Majesty's Service. The difficulty of succeeding in London induced me to visit some places in the country and after some years spent in Newcastle, Leeds &c: I was chosen Organist at Hallifax in Yorkshire. during all this time, tho' it afforded not much leisure for study I had not forgot my former plan but had given all my leisure hours to the study of languages. After I had improved myself sufficiently in english I soon acquired the italian, which I looked upon as necessary to my business; I proceed[ed] next to latin & having also made a considerable progress[a] in that language I made an attempt of the greek, but soon after dropt the pursuit of that as leading me too far from my other favourite studies by taking up too much of my leisure. The theory of Music being connected with mathematics had induced me very early to read in germany all what had been wrote upon the subject of Harmony; & when, not long after my arrival in England the valuable Book of D^{r}. Smith's Harmonics[2] came int[o] my hands I perceived my ignorance & had recourse to other authors for imformation by which means I was drawn on from one branch of the Mathematics to another. In the Year 1766 I was removed from Hallifax to Bath where I became Organist of the Octagon chapel. My Situation proved a very profitable one, as I soon fell into all the public business of the Concerts, the Rooms, the Theatre & the Oratories, besides many Schollars & private Concerts. The great run of business far from lessening my attachment to study encreased it, so that many times after a fatiguing day of 14 or 16 hours spent in my vocation I retired at night with the greatest avidity to <u>unbend the mind</u> (if it may be so called) with a few propositions in Maclaurin's Fluxions[3] or other Books of that sort. Among other mathematical Subjects optics and Astronomy came in turn & when I read of the many charming discoveries that had been made by means of the telescope I was so delig⟨h⟩ted with the subject that I wished to see the heave⟨n⟩s & Planets with my own eyes thro' one of those instruments Accordingly I hired a 2 f^{t}. gregorian Reflector this being the best instrument the town would afford. The satisfaction I received determined me to furnish myself with a capital telesc & ignorant of the value of these instruments I desired a 5 f^{t}. reflect^{r}[4] to be made for me. The person who was employed to procure it received advice of the terms & thought proper to acquaint me with them. The price tho' moderate appeared to me so extravagant that I formed the resolution to make myself one; as, not aware of the difficulty, it appeared to me from some former mechanical attempts that with the assistance of

[a] *altered from*: compleated myself.

the directions given in Dr Smiths optics[5] I should be able, in time, to accomplish such a work. In the pursuit of this laborious, but to me delightful undertaking I persisted for some years with unwearied assiduity till to my infinite satisfaction I saw Saturn in the year 1774 thro' a 5 foot Newton⟨^n.⟩ reflect of my own making. This success proved such an encouragement that I proceeded to larger instruments, & my Situation in business furnishing me amply with the means of carr⟨y⟩ing on my improvements, I soon made a 7 f^t. a 10 f^t. & a 20 feet reflector of the same construction. I persisted with such obstinacy in compleating the parabolical figure of a 7 f^t. telescope that I made above 200 object specula, till at length[a] I obtained one that would bear any power I could apply to it.[6] All this while I continued my astronomical observations & nothing seemed now wanting to compleat my felicity than suficient time to enjoy my telescopes to which I was so much attached that I used frequently to run from the Harpsichord at the Theatre to look at the stars during the time of an act & return to the next Music. To this perseverance at length was owing the Discovery of the Ge[orgium] Sid[us] which happened on the 13 of March 1781.[7] It has generally been supposed that it was a lucky accident that brought this star to my view. This is an evident mistake. In the regular manner I examined every star of the heavens not only of that magnitude but many far inferior it was that night its turn to be discovered. I had gradually perused the great Volume of the Author of Nature & was now come to the page which contained a seventh Planet. Had business preventd me that evening, I must have found it the next, and the goodness of my telescope was such that I perceived its[b] visible planetary disk as soon as I looked at it; And by the application of my Microm^r determined its mot[ion] in a few hours. The same year I had the honnour of be[in]g admitted a fellow of the RS. and received the Gold Medal for the discovery of the Planet. In the Spring of the next year His Majesty wished to see my teles. &, ever ready to encourage Arts & Sc^s. was pleased with his Roy⟨l⟩ Goodness [to] take me from my former employm^t. that I might devote myself intirely to Ast. In Jan 1781 I began to make a 30ft Newt[onian] Ref[lector] & soon after twice cast a Spec[ulum] of 36 inch Dia[mete]^r but neither of them succeeded at that time. The interruption also of my removal from Bath put a temporary stop to the work. My present Situation made it prudent to undertake an instrum^t of a smaller construction which I have happily executed & the result of one years observat[ion] with it have already been partly comunicated to the RS. I am however still in

---

[a] *altered from*: last.

[b] *altered from*: the.

hopes soon to be able to resume & perhaps to go beyond my former attempt of a larger inst[rument].

Notes

For William Herschel see note to Letter 21. This is (a copy of) the account of Herschel's life Hutton requested in Letter 25; it was based on previous letters to Lalande and to G.C. Lichtenberg from 1782 and 1783: see Gingerich, 'William Herschel's 1784 Autobiography', 77. The letter that accompanied it does not survive; the printing of this text in the *European Magazine* is addressed in Letters 27, 28, and 29.

[1]The allusion was to events during the Seven Years' War. Herschel, like his father, played in the band of the Hanoverian footguards; the guards suffered defeat in the battle of Hastenbeck in July 1757 and Herschel, after fleeing to Hanover, was in danger of being pressed into service as a soldier; thus he fled again, this time to England: ODNB.

[2]Smith, *Harmonics*.

[3]MacLaurin, *Fluxions*, one of the last and most important British textbooks on the calculus from the period of Newton and his circle.

[4]The lengths given in the text are the focal lengths of the principal mirrors in each of the telescopes described: not, as in modern usage, their diameters.

[5]Smith, *Opticks*, by the same author as the 'Harmonics' mentioned earlier in this letter.

[6]That is, the image produced by the principal mirror remained sharp in an eyepiece of any power Herschel applied to it.

[7]Compare Letter 21, where the planet now known as Uranus was called 'the Planet Herschel'.

Letter 27.
CHARLES HUTTON to WILLIAM HERSCHEL, 18 November 1784

London, Royal Astronomical Society, William Herschel Correspondence 13.H.37: ALS (our source).

Also printed in Gingerich, 'William Herschel's 1784 Autobiography', 80–1.

Woolwich, Nov. 18, 1784.

Dear Sir,

In haste I have just time to inform you of the safe receipt of your Letter with the Memorandums,[1] so modestly & fairly detailed, which I hope will do you honour. I have mixed the whole up among so⟨me⟩ remarks of my own,[2] & shall carry them up to town with me to day, & deliver the paper[a] in in [*sic*] a state I hope fit to be printed, without any farther interference of any one else, & I hope it will arrive before they shall have printed any other less perfect, & less just acco[t]. of their own. You may depend on all possible prudence being used with respect to your communicating, & they shall be welcome to give me up if[b] they please, as I shall always esteem it an honour to have done justice in any degree to so deserving a character. And am

Dear Sir

Your sincere friend

& hble Servant

Cha Hutton.

William Herschel Esq.
Datchet
Windsor

Notes

For William Herschel see note to Letter 21. This letter is a reply to Letter 26 (or rather to the letter that accompanied that document); the same matter is addressed further in Letters 28 and 29.

[1]Letter 26.

[2]The reference is to Letter 26; various changes were made for printing it as Herschel, 'William Herschel'. First person was changed to third person throughout and orthography and paragraphing were emended; there were added one introductory and one concluding sentence, an admiring reference to a solution Herschel contributed to the *Ladies' Diary* in 1780, and a list of Herschel's papers in the *Philosophical Transactions*.

Letter 28.
CHARLES HUTTON to WILLIAM HERSCHEL, 31 January 1785

London, Royal Astronomical Society, William Herschel Correspondence 13.H.38: ALS (our source).

Also printed in Gingerich, 'William Herschel's 1784 Autobiography', 81.

---

[a] *altered from*: it.

[b] *altered from*: as.

Woolwich 31 Jan[y]. 1785

Dear Sir

Your favour[1] did not arrive at Woolwich according as might[a] be expected from the Date of it, which was monday morning, which was the 24th, whereas the Letter was not put into the General post till the 25th, & from thence into the Penny post the 26th, as you will see by the Cover here inclosed. And as I received it only on the Wednesday Evening, I did not write, as I should be in town on the Thursday myself as soon as my letter could be, & if you should be in town I expected to meet you at the R. Soc[y].[2] but not finding you there I went the next day to Jack's Coffee house in Dean Street Soho to enquire for you, when the keeper informed me you had not been there for several weeks before, so that I apprehend some mistake in the place. Before I left town I corrected the proof sheet of the Anecdotes in the European Magazine, & introduced the little alteration you desired.[3] It will be published tomorrow. The Editor promised to send me a few proof impressions of the head in a few days, which I shall send to you as soon as I receive them. The Gent[n]. of the Mathem[l]. Club[4] will be always glad to see you, & desire you will come as often as you can. We met last Friday, & you will observe we meet every-other Friday a little before 4, so from that Epoch you can always calculate the days. I am,

Dear Sir
Your obed[t] Hble Servant
Cha. Hutton.

W[m] Herschel Esq
Datchet
Windsor

Notes

For William Herschel see note to Letter 21. This letter continues the discussion of the printing of Letter 26 in the *European Magazine*, begun in Letter 27.

[1]Herschel's letter to Hutton, evidently dated Monday 24 January, is not known to survive.

[2]This seems to imply that Hutton was still attending some meetings of the Royal Society at this date, despite what is asserted in Wardhaugh, 'Charles Hutton and the "Dissensions"'.

[3]Gingerich (p. 81) identifies the 'little alteration' as the transformation of

[a] *altered from*: to the.

His Majesty wished to see my teles.

into

his Majesty having been informed by the President of the Royal Society of the discovery of the New Planet, wished to see his telescope

(p. 2 of the printed text). This is indeed the only change that seems at all likely to have exercised Herschel himself.

[4]On the mathematical or 'Friday' Club, see note to Letter 20.

Letter 29. CHARLES HUTTON to WILLIAM HERSCHEL, 15 February 1785

London, Royal Astronomical Society, William Herschel Correspondence 13.H.39: ALS.

Woolwich, Feb. 15, 1785.

Dear Sir

I hoped to have had the pleasure of meeting you at the Mitre on Friday last,[1] and to have delivered to you a couple of fine proof impressions of the head, which I had put in my pocket for that purpose. Being disappointed of your company, I left them for you with the Master of Jack's Coffee house in Dean Street who will deliver them to you.

Dear Sir
Your most hbl Ser[t]
Cha. H[utton][a]

Note

For William Herschel see note to Letter 21. This letter concludes the discussion of the printing of Letter 26, begun in Letters 27 and 28.

[1]On the mathematical or 'Friday' Club, see note to Letter 20.

---

[a]The paper is torn.

Letter 30.
Lewis Evans to Charles Hutton,
25 April 1785

Oxford, Museum of the History of Science, MS Evans 31: draft.

Compton
near Farringdon
Berks
25th. April 1785.

Sir.

I send you this year a very small tribute indeed to the Ladies' Diary: my time having[a] been so occupied in other professional matters [*long deletion obliterated*]. I remain

Sir your devoted humb. Servt.
L: Evans

Note

For Lewis Evans see note to Letter 17.

Letter 31.
Charles Hutton to Nevil Maskelyne,
27 June 1785

Cambridge, University Library, RGO 4/187/11, fols. 1–2: ALS.

The two papers enclosed with the letter do not survive. The letter is covered with Maskelyne's calculations and sketches.

---

[a] *altered from*: has.

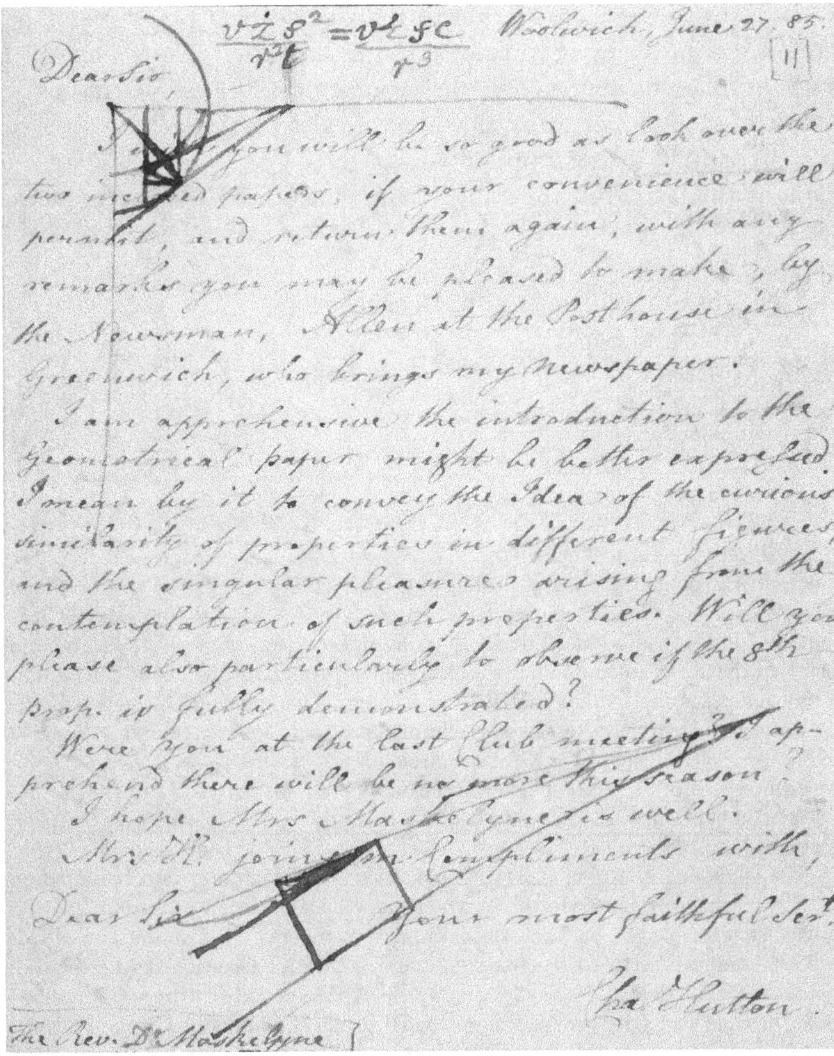
$\frac{v\dot{z}s^2}{r^2t} = \frac{v\dot{z}sc}{r^3}$ Woolwich, June 27. 85.

[11]

Dear Sir,

I wish you will be so good as look over the two inclosed papers, if your convenience will permit, and return them again, with any remarks you may be pleased to make, by the Newsman, Allen at the Posthouse in Greenwich, who brings my Newspaper.

I am apprehensive the introduction to the Geometrical paper might be better expressed. I mean by it to convey the Idea of the curious similarity of properties in different figures, and the singular pleasures arising from the contemplation of such properties. Will you please also particularly to observe if the 8th prop. is fully demonstrated?

Were you at the last Club meeting? I apprehend there will be no more this season.

I hope Mrs Maskelyne is well.

Mrs H. joins in Compliments with, Dear Sir, Your most faithful Sert.

Cha. Hutton.

The Rev. Dr Maskelyne

Figure 2: Letter 31: Charles Hutton to Nevil Maskelyne, 27 June 1785. Cambridge University Library, RGO 4/187/11, fol. 1r. Reproduced by kind permission of the Syndics of Cambridge University Library.

Woolwich, June 27. 85.

Dear Sir,

I wish you will be so good as look over the two inclosed papers, if your convenience will permit, and return them again, with any remarks you may be pleased to make, by the Newsman, Allen at the Posthouse in Greenwich, who brings my newspaper.

I am apprehensive the introduction to the Geometrical paper might be better expressed. I mean by it to convey the Idea of the curious similarity of properties in different figures, and the singular pleasure arising from the contemplation of such properties. Will you please also particularly to observe if the 8th prop. is fully demonstrated?[1]

Were you at the last Club[2] meeting? I apprehend there will be no more this season?

I hope Mrs Maskelyne is well.

Mrs H.[3] joins in Compliments with, Dear Sir

Your most faithful Ser^{t}.

Cha. Hutton.

The Rev. Dr Maskelyne

Notes

For Nevil Maskelyne see note to Letter 13.

[1]The 'Geometrical paper' was quite probably a draft of Tract VII: 'Of the Common Sections of the Sphere and Cone ...' in Hutton, *Tracts* (1786), 83–92 (there dated 29 July 1785). As published, it was introduced by three paragraphs reflecting on the 'pleasure and delight' arising from 'the beautiful analogy which reigns through all the objects of similar enquiry' in mathematics, and in geometry in particular. The eighth proposition in the published paper was this (p. 89): 'Of the two common sections of a sphere and an oblique cone, if the one be a circle, the other will be a circle also.'

[2]This was quite possibly a reference to the 'Friday Club'; see note to Letter 20.

[3]This referred not to Hutton's first wife Isabella (see note to Letter 16), who had died near Newcastle in May 1785, but to Margaret Ord (*c.* 1752–1817). It is not known whether they met in Newcastle or in London (it is possible she was the Margaret Ord baptized on 9 July 1752 at Norham, Northumberland, in which case the connection presumably dated from Hutton's Newcastle days), or when she became a member of his London household. In January 1778 she gave birth to Hutton's daughter Charlotte (see note to Letter 67), and a few weeks after the death of Isabella, on 23 July 1785, she and Hutton were married at St Dunstan in the West (the parish church for Hutton's chambers in Clement's Inn): Mackenzie, *Historical Account*, 560; information from IGI and FR; see also Letters 113 and 116. The present passage raises the possibility that Hutton had already introduced Margaret to his London friends as 'Mrs Hutton'.

Letter 32.
DAVID STEUART ERSKINE, EARL of BUCHAN to CHARLES HUTTON, 26 January 1786

Edinburgh, National Library of Scotland, MS 967, fol. 291: ALS.

Sender reported in endorsement on cover: 'Earl Buchan'.

Letter 32. David Steuart Erskine, Earl of Buchan

Great Pulteney Street N°. 32.
Thursday 26th. January 1786.

Sir,

His Majesty having returned to me with a very flattering compliment on its subject the account of the life writings & Inventions of John Napier of Merchiston,[1] I am now ready to commit to your obliging care what is ready of that Tract for the press, and before the printing can be far advanced I shall be in readiness for the Appendices.[2]

Were it convenient for you to come to Town for a day or two to settle this business there is an appartment for you here in my House. If not I shall wait upon you at Woolwich notwithstanding much pressing business which obliges me to be much on the watch here to meet those with whom I am to transact it. As this tract of Napiers is of a very interesting nature to the literary world all over Europe I have had thoughts of printing it in Latin as well as in our language,[3] and as to the profits of the sale I mean to apply them to the use of my learned coadjutor M^{r}. Walter Minto a Man of great abilities whose modesty has hid him too much from the publick Eye;[4]

I am Learned Sir,
with great regard,
your very much obliged
Humble Servt.
Buchan.

To
D^{r}. Charles Hutton
Professor of the Mathematics
Woolwich Academy

Notes

David Steuart Erskine, eleventh earl of Buchan (1742–1829), antiquary and political reformer, was the son of parents who had both studied mathematics with Colin MacLaurin, and a Fellow of the Royal Society from 1764. After his retirement from politics in 1780 he concentrated on antiquarianism, founding the Society of Antiquaries of Scotland in that same year. It is not known how or when he became acquainted with Hutton: ODNB.

[1]John Napier (1550–1617), inventor, theologian, and mathematician, is best known as the inventor of logarithms, announced in Napier, *Descriptio*: ODNB.

[2]Buchan and Walter Minto (for whom see note 4) had in 1778 published an account of Napier's life, dedicated to the King. This passage seems unmistakably to envisage a new printing of that work, but in fact it was merely reissued in 1787 with a new title page and list of errata, the London bookseller being now Robinson rather than Murray. Presumably the stock of unsold copies was found too large to justify a fresh printing, whatever alterations or new matter Buchan had prepared.

[3]The 'Tract' in question was most likely Napier, *Descriptio*, or Napier and Briggs, *Constructio*. In fact no separate new edition of any work by Napier was printed in London in the 1780s or 90s. Maseres, *Scriptores Logarithmici*, vol. 6, pp. 475–624, however, comprised a re-edition of Napier, *Descriptio*, and although neither Hutton nor Buchan was mentioned there, it may be relevant that Playfair believed Hutton to be involved with the *Scriptores*: see Letter 39.

[4]Walter Minto (1753–96) was born in Scotland and held the mathematical chair at the University of Edinburgh for a period; he moved to America in 1786 and became professor of mathematics and natural philosophy at the College of New Jersey. He was the coauthor of Erskine and Minto, *John Napier*: see Eisenhart, 'Walter Minto and the Earl of Buchan'.

Letter 33. LEWIS EVANS to CHARLES HUTTON, 24 April 1786

Oxford, Museum of the History of Science, MS Evans 31: draft.

Compton
April 24th. 1786

Sir.

The following solutions are totally at your service and I hope you will ⟨find⟩ them to be right.

I am Sir
your very humble Servt.
L: Evans

...

I must confess that the solution ⟨I have given⟩ to question the 6th. is nothing more than groping at one, because I have not ascertained the relation of b to a and the answers are ⟨therefore⟩ accidentally true.[1]

I will beg leave to request an insertion of the following new question.

...

Note

For Lewis Evans see note to Letter 17.

[1]Question 6 in the 1786 *Ladies' Diary* was 'To find two numbers so, that their difference, the difference of their squares, and the difference of their cubes, may be all square numbers.'

Letter 34.
LEWIS EVANS to CHARLES HUTTON, 1786–7

Oxford, Museum of the History of Science, MS Evans 31: fragment of draft.
Date and recipient inferred from the position of this fragment in the bound volume of Evans' letters to Hutton.

Sir

Prompted by the fair opportunity presenting itself for rendering my situation more agreeable I avail myself of every essential support that may tend to my success and have taken the liberty to solicit the favour of your kind aid towards effecting that end, flattering myself that the request I have to make will meet with your willing ac-[1]

[*The remainder of the page has been cut off and the end of the letter, if any, is absent. The following page resumes with Evans' solutions to the questions in the 1787 Ladies' Diary, which were presumably sent in as usual during April of that year; their covering letter is missing.*]

Note

For Lewis Evans see note to Letter 17.

[1]The change in Evans' address—Hungerford, from April 1788—together with the evidence of Letter 41, indicates that he successfully applied for a position at the school in Hungerford between April 1786 and April 1788. It is not clear whether a version of this letter was sent to Hutton, that is whether Evans sought—or received—Hutton's support for his application.

Letter 35.
CHARLES HUTTON to JOHN WHITEHURST, 16 August 1786

Birmingham, Archives, Heritage and Photography Service, MS 3782/12/31/97: ALS (our source).
Extract printed in *The Lunar Society*, 30.

John Whitehurst forwarded this letter to Matthew Boulton: the inside cover bears the address *Matthew Boulton Junr. Esqr / Soho / Birmingham*, and the letter has been re-folded, re-sealed, and franked accordingly.

Shooters Hill, Aug. 16, 1786.[a]

Dear Sir,

I know not how your letter, dated the 13th, has been delayed, but it only arrived here this day at Noon, so could not give you an ans^r sooner. I thank Mr Bolton for his good opinion of me, in offering to place his son in my house.[1] You know I do not take boarders. But in the present instance, to oblige so worthy a friend, if he do not think 100 Guin^s. P^r An. too much, I shall endeavour to give him satisfaction.

No, Sir, Dr R.[b] has not informed me about the comparison of your Measure with any standard. I am glad to find the result so exact, for the length of the Second Pendulum, viz. 39.129.[2] I always use the number $39\frac{1}{8}$, or, which is the same thing, 39.125; which you see is very near your number, above given.

I am,
D^r S^r
Your obed^t Servt
Cha. Hutton.

Mr Whitehurst
Bolt Court
Fleet Street

Notes

John Whitehurst (1713–88), maker of clocks and scientific instruments, and geologist, was a member of the Lunar Society of Birmingham as well as a Fellow of the Royal Society; Hutton would edit his *Works* in 1792: ODNB.

[1]Matthew Boulton (1728–1809), manufacturer and entrepreneur, was, like Whitehurst, both a Fellow of the Royal Society and a member of the Lunar Society of Birmingham. He is best known for his partnership with James Watt and the manufacture and marketing of improved steam engines in the period from the mid-1770s.

His son was Matthew Robinson Boulton (1770–1842), who became a director of Soho Foundry, Soho Mint, and Soho Manufactory after his father's death: ODNB. Boulton in fact took his son with him to France later in 1786; I have seen no reference to his boarding with Hutton outside this letter, indicating perhaps that Hutton's offer was not taken up: see Smiles, *Boulton and Watt*, 403–8, and Dickinson, *Matthew Boulton*, 164–6.

---

[a]Added to address on cover: Wednesday evening.
[b]or *B*.

[2]Whitehurst had been a judge in 1774 for the award offered by the Society for the Encouragement of Arts, Manufactures, and Commerce for the best solution to the problem of producing a natural standard of length, and subsequently developed the idea of the winning entry in Whitehurst, *An Attempt*. 'Dr. R.' may possibly have been Boulton's patron the ironmaster John Roebuck (1718–94).

Letter 36. CHARLES HUTTON to JOHN ROBISON, ?late 1786

Printed in Hutton, 'The Expansive Force of freezing Water', 23–4. The letter was read to the Royal Society of Edinburgh, of which Robison was General Secretary, on 6 November 1786: see Letter 39.

SIR,

THE following is an extract of a letter to me from Major EDWARD WILLIAMS, of the Royal Artillery,[1] a learned man, and of great professional merit. Being at Quebec in some very cold winters, among various other ingenious experiments, it occurred to him to try the force of congelation in some of the iron bomb-shells, which are usually fired out of mortars in the practice of artillery; by filling the cavity of the shell with water, and then, having plugged up the fuze-hole, exposing it to the cold to freeze the water, in order to find whether the expansion of the ice would be capable of bursting the shell.

THE dimensions of the 13 inch shell are as follow:

	Inches
Outer diameter of the shell,	12.8
Inner, or diameter of the cavity,	9.1
Thickness of metal at the fuze-hole,	1.5
Ditto at the bottom or opposite part,	2.2
Diameter of the fuze-hole,	1.7

And the dimensions are similar in the other shells. The fuze-hole is conical, the opposite sides of which, when produced, meet at the extremity of the diameter, which passes through the middle of the hole.

HE found, that the iron plug could hardly ever be driven so firmly into the fuze-hole as to resist the force of the expansion of the ice, which pushed it out with great velocity, and a bolt or cylinder of ice, of a considerable length, immediately shot up from the hole. But when the plug was fixed in with springs, which laid hold of the inside of the cavity, so that the plug could not possibly be pushed out, the force of expansion then split the shell, and a fin or plate of ice shot out quite around.

[*A long extract from Williams's paper—in the form of a letter to Hutton—follows, describing experiments done in 1784 and 1785, with concluding 'remarks' by Hutton.*]

Note

John Robison (1739–1805), natural philosopher and inventor, was at this time general secretary to the Royal Society of Edinburgh: ODNB.

[1]For Edward Williams see notes to Letter 9.

Letter 37. CHARLES HUTTON to CHARLES LENNOX, DUKE of RICHMOND, 24 August 1787

Dedicatory epistle printed in Hutton, *Conic Sections*, v–vi.

Royal Mil. Acad.
Aug. 24, 1787.

TO HIS GRACE
CHARLES,
DUKE OF RICHMOND, LENNOX, AND AUBIGNY,
&C. &C. &C.
MASTER GENERAL OF THE ORDNANCE.

MY LORD,

SHOULD this small performance continue to prove useful to the Institution for which it was composed, or become acceptable to the public; its existence, as well as its publication, must be ascribed to YOUR GRACE.

But, my LORD, the particular injunctions to avoid panegyric, under which permission for this dedication was granted, preclude me from explaining how the advanced state of learning in the Royal Military Academy rendered such a work necessary: since this would be to enumerate the means by which, under Your Auspices, this institution has attained a degree of perfection, which perhaps few public ones have ever equalled, none certainly have exceeded.

To YOUR GRACE therefore, with all Respect, this work is most humbly dedicated by,

MY LORD,
YOUR GRACE'S
most obedient,
most devoted, and
most humble servant,
CHARLES HUTTON.

Notes

Charles Lennox, third duke of Richmond, third duke of Lennox, and duke of Aubigny in the French nobility (1735–1806), politician, had reached the rank of colonel in 1758 and was, after his retirement from active service in 1760, Master-General of the Ordnance from 1782 to 1783 and again from 1784 to 1795. As such he was the head of the Board of Ordnance which oversaw the Royal Military Academy at Woolwich where Hutton taught: ODNB.

Letter 38.
CHARLES HUTTON to HUGH PERCY, 1788

Dedicatory epistle printed in Hutton, *Mensuration* (2nd edition, 1788), iii–iv.

TO
THE MOST NOBLE AND PUISSANT PRINCE,
HUGH,
DUKE AND EARL OF NORTHUMBERLAND,
EARL AND BARON PERCY, LUCY, POYNINGS, FITZPAYNE, BRYAN, LATIMER, AND WARKWORTH, AND BARONET;
LORD LIEUTENANT AND CUSTOS ROTULORUM OF THE COUNTY OF NORTHUMBERLAND, AND OF THE TOWN AND COUNTY OF THE TOWN OF NEWCASTLE UPON TYNE; VICE ADMIRAL OF THE COUNTY OF NORTHUMBERLAND; AND LIEUTENANT GENERAL OF HIS MAJESTY'S FORCES, COLONEL OF THE SECOND TROOP OF HORSE GRENADIER GUARDS, &C. &C. &C:

MY LORD,

IF the high honour this Work formerly received in the countenance of Your Grace's illustrious Father,[1] had not particularly encouraged the Author's presumption, in thus seeking the protection of his noble Representative, he is convinced that he could not easily have found among the great and elevated, a character under whose auspices he should more earnestly have wished to present it to the world in this its improved state.

The great progress of all the useful arts and sciences in this country, since the happy æra of the accession of our august Sovereign, must be ascribed, next to his benign and munificent influence, to that eminent countenance which some of the greatest characters have shewn, by their personal example in the cultivation of them, not less than by their patronage and protection.

To the hereditary fame of Your Most Noble Progenitors, Your Grace's personal bravery and military talents have added great and distinguished lustre.[2] The public confidence and esteem which necessarily follow such accomplishments, stamp a value on every thing, however inconsiderable, which is honoured with Your Grace's patronage: And the accurate judgment which Your Grace is known to possess on all subjects of extensive practical utility, makes it the wish of such as cultivate them, ardently to seek Your Grace's protection.

With these impressions, I presume to lay the following performance at Your Grace's feet; and am, with the profoundest respect,

My Lord,

Your Grace's

most obedient, and

most devoted humble servant,

CHARLES HUTTON.

Notes

Hugh Percy, second duke of Northumberland (1742–1817), army officer and politician, was the eldest son of Hugh Percy, first duke of Northumberland, for whom see Letter 1: ODNB.

[1]See Letter 1, the dedicatory epistle to the first (1770) edition of Hutton's *Mensuration*, addressed to the previous Duke of Northumberland.

[2]Hugh Percy had achieved distinction as an army officer in the early stages of the American War of Independence, from 1775 to 1777.

Letter 39.
JOHN PLAYFAIR to CHARLES HUTTON, 21 April 1788

London, Wellcome Collection, MS 7430 no. 38: ALS.

Sender reported in endorsement (in Hutton's hand): *John Playfair / 21 April 1788.*

Edin^{r}. Ap. 21st 1788

Dear Sir

I know not to what accident it may have been owen that I did not receive your letter dated the 31st Octr. 1787, and the books that accompanied it, till a few Days ago. I am now happy to thank you for them both, and as you will naturally accuse me of great negligence in not answering your letter, and[a] especially the commission you give me relative to a little work of James Gregory, I have not waited for a Frank but am to take the liberty of putting you to the expence of a Postage tho' I have not any thing very material perhaps to send you.

By the first opportunity I can find I will send you the book printed at St. Andrews in 1672, under the signature of Pat. Mather &c.[1] It contains[b] beside the great and new Art of weighing Vanity; An Examination of M. Sinclar's Hydrostaticks; Examen Artis Novæ & Magnæ Georgii Sinclari; Tyrociniorum Mathematicorum examen; to all which are added Tentaminæ quædam Geometricæ de Motu penduli, et Projectorum. These treatises are all very short and compose a small Duodecimo. The last is the most valuable, for the controversial pieces, tho' their author be[c] in the right, are conducted with very little liberality or good manners, yet they have now & then some little approach to Wit, & the first is in English remarkably good for the time and place when it was written. You will judge of all this when you receive the Book, which shall be as soon as possible.

Dr Gregory of this University, Prof. of the Theory of Physic, and a worthy descendant from so illustrious an ancestor as James Gregory of Aberdeen, tells me that he has a great many Papers of David Gregory's[d] which have never been looked into & promises me a sight of them all.[2] It is probable I think that they may contain either something of James Gregory's own, or some anecdotes of him that you would wish to know. I shall look thro' them with great care, and transmit you whatever I find of any consequence. your letter however has been so long on its way that I fear very much all this will come too late for your Work. Be so good as let me know whether it will or not. If you would wish to see Sinclar's Books,[3] I will endeavour to find them out likewise. Indeed it would make me happy to contribute in any thing to a Work so interesting and important as that in which you and M^{r}. Maseres are engaged, and I cannot but congratulate my Country-Man Dr G. as well as the other authors on Logarithms, on having their works [il]lustrated by the learning and ability of two [s]uch Editors.[4]

---

[a] *altered from*: And.

[b] *deleted*: the great.

[c] *altered from*: being.

[d] *altered from*: Gregorie's.

It must have arisen from the[a] indisposition of Mr Prof Robison[5] and the languor occasioned by it, that you was not informed both of the safe arrival of your communications here, & of their being very favourably [re]ceived by the Society. The 1st viz the account [of] the Experiments on Congelation was read Novr. 6th. 1786: the 2^d on the resistance of the air on Jan. 1. 1787. They will both appear in the 2^d vol. of their Transactions now preparing for the press, and you may be assured that we are extremely thankful to you for such valuable communications.[6] I am very glad to think the expression of ⟨the⟩ airs resistance is likely to be so simple as you mention in y^r letter.[7]

I am much pleased with your Conicks,[8] so far as I have yet considered them, and expect much advantage both from them and from the collection of Problems at the end.

I must intreat you to present my respectful compliments to Captn Dorset,[9] and to Dr Maskelyne.

I have left myself no room to inquire about your opinion of our Transactions, nor to thank you for the good opinion you express of my Paper.[10] But I hope soon to hear from you, and at any rate shall write at more length by the Book, I am to send, which will probably be brought you by Mr Prof. Stewart[11] who will pass thro' London on his way to the continent in a week or two.

Doctor Hutton
Academy Woolwich
By London

Notes

For John Playfair see note to Letter 19.

[1]Playfair described 'Mathers', *The great and new art of weighing vanity*, usually attributed to James Gregory (1638–75). Hutton retained an interest in this particular work to the end of his life; in the introduction to Maseres and Babbage, *Scriptores optici* (1823), Charles Babbage reported that 'The Great Art of Weighing Vanity' was included in that volume on Hutton's advice (p. 5).

[2]James Gregory (1753–1821) was Professor of the Institutes of Medicine at Edinburgh from 1776; his great-grandfather was the mathematician James Gregory (1638–75), whose nephew was the mathematician, astronomer, and Savilian professor David Gregory (1659–1708): all ODNB.

[3]Referring to the works of George Sinclair (d. 1696), 'examined' in 'Mathers', *The great and new art of weighing vanity*, mentioned in note 1.

[4]Playfair apparently referred to Maseres, *Scriptores Logarithmici*, usually attributed to Francis Maseres alone (for whom see the note to Letter 7). As far as I can tell, that

[a] *deleted*: ?dist....

work did not make use of unpublished or biographical material about Gregory; nor did Hutton's biographies of James and David Gregory for his *Dictionary* (vol. 1, pp. 600–6) make evident use of any papers from Playfair.

[5]For John Robison see note to Letter 36.

[6]Hutton, 'The Expansive Force of freezing Water' (for which see also Letter 36) and Hutton, 'The true resistance of the air'.

[7]Hutton's experiments on ballistics and the resistance of the air to projectiles took place in 1783–5, 1787–9, and 1791, and his ideas about their interpretation evolved greatly up to the publication of his 1812 *Tracts*; that he did not find a workable law describing the resistance of the air was perhaps the most important failure of his scientific career. At the end of 1786, based on work at low velocities, he believed that resistance varied nearly in proportion to the surface area of a projectile and a little faster than the square of its velocity (Hutton, *Tracts* (1812), vol. 3, p. 190); it seems likely that this was the model he had mentioned to Playfair. By 1795, further experiments would yield 'a complete series of the resistances of the air' for various velocities from 0 to 2000 feet per second (Hutton, *Dictionary*, vol. 1, p. 185) and by 1812 Hutton had come to despair of finding an algebraic model that would fit his data: see in Hutton, *Tracts* (1812), vol. 3, the tables at p. 88, the graphs at p. 108, the analysis and formulae at, for example, pp. 222–9, and the admissions of failure at pp. 128–9 and pp. 256–7 ('no hope remains of the practicability, if even of the possibility, of obtaining a correct and scientific solution of the general problem in that way').

[8]Hutton, *Conic Sections*.

[9]Michael Dorsett or Dorset had been commissioned as a second lieutenant in the Royal Artillery in 1771, but later transferred to the Sussex Militia, where he became a captain: Kane, *List of Officers*, 14; Bell, *Bell's List*, 48; ODNB *s.v.* 'Dorset [née Turner], Catherine Ann (bap. 1753, d. in or after 1816), children's writer' (his wife). For Nevil Maskelyne see note to Letter 13.

[10]The reference was presumably to the only volume of the *Transactions of the Royal Society of Edinburgh* that had so far appeared: volume 1, dated 1788. Playfair, 'On the Causes' appeared therein, and was probably the paper referred to here.

[11]For Dugald Stewart see note to Letter 19.

Letter 40.
LEWIS EVANS to CHARLES HUTTON, 28 April 1788

Oxford, Museum of the History of Science, MS Evans 31: draft.

Hungerford
April 28th.
1788

Dear Sir

Since I sent you my packet of solutions[1] I have discovered a trifling mistake in my solution of the 5th. Question. I shall now therefore give it you in a different form and[a] something clearer to those of less discernment[b] than D[r]. Hutton.

...

Should you have any[c] intention to[d] publish another[e] Edition of your Arithmetic[2] I could communicate to you what in my humble opinion would be emendations of some of the general Rules which came of late more immediately under my practice.

When I was at Greenwich Observatory I was highly entertained & delighted with the several Instruments, which without doubt are as accurate as hands cd[f] make them, and among the rest the Time-piece standing by the mural Quadrant attracted my notice, the pendulum of which I observed was of the gridiron kind ⟨of⟩ the nature of which I have some faint Idea but am[g] ignorant of its construction and application to the Palate[h]. Notwithstanding, I am anticipating the pleasure of having one day or other a tolerably accurate Chronometer in the hope of being favoured by your kind assistance with the construction of this pendulum, which I find obviates in a great measure the irregularities arising from the expansion and contraction of the wire in common pendulums by the effects of heat and cold. Doctor Maskelyne very much commended[i] Graham's barometrical pendulum[3] which I fear will be too expensive for me to make choice of were I acquainted with its form.

In a short time you will be fully engaged in the arrangement of the Diary. If you mean to publish another supplement to it,[4] perhaps it may be a recommendation to its sale[j], were you to divide it into three equal parts; one part appropriated for the supernumerary[k] answers of Enigmas, & Charrands; another for the answers to the philosophical queries; and the third for selected mathematical solutions. So that no new Enigmas or Queries be proposed at all in it. ⟨This I offer as my opinion only with respect to its general reception⟩ But this Division of the work, I may venture to aver,

---

[a] *deleted:* perhaps.
[b] *altered from:* less ?conversant.
[c] *altered from:* inform me of your.
[d] *altered from:* of.
[e] *deleted:* another.
[f] *altered from:* can.
[g] *altered from:* am not ?totally.
[h] *altered from:* Verge.
[i] *deleted:* M[r].
[j] *altered from:* better reception.
[k] *altered from:* additional.

will not meet the entire approbation of the Bulk of Mathematicians. They would rejoice in the ⟨present⟩ opportunity, under the superintendance of so able a conductor, of having as many as may be of those elegant solutions, the productions of profound Geniuses, transmitted to Posterity, what would otherwise perhaps be consigned to oblivion. I am Dear Sir

your ever obliged

humble servant.

L.E.

Having sent this under cover to a Brother of mine, who is at present in London, probably, if he is not particularly streightened in Time he may personally deliver this ⟨to your hands⟩, while it will afford him great pleasure in paying his respects to so good a Friend of his Brother's.

Notes

For Lewis Evans see note to Letter 17. This was the first letter in which he mentioned Hutton's name, although he seems to have received a personal reply to his letters to the editor of the *Ladies' Diary* as early as 1783 (see Letter 23).

[1]Evans' covering letter to his solutions for the 1788 *Ladies Diary* is not extant: see note to Letter 34.

[2]Hutton's 1764 *Guide*, of which the eighth edition appeared in 1788 and the ninth in 1792.

[3]Evans referred to the temperature-compensated pendulum invented by the horologist and scientific instrument-maker George Graham (*c.* 1673–1751), which used a glass jar of mercury for a bob: ODNB, MPHE. For Nevil Maskelyne see note to Letter 13.

[4]From 1788 Hutton produced annually, as well as the *Ladies' Diary*, a *Supplement* or *Diary Companion*, containing additional material 'for which there was not room in the Diary itself'.

Letter 41.
LEWIS EVANS to CHARLES HUTTON, 15 July 1788

Oxford, Museum of the History of Science, MS Evans 31: draft.

Hungerford
July 15th
1788

Dear Sir

A few days ago upon looking over my solution to the 11th. Question I found that I had taken h (= the height of the cone described by the staff's center of oscillation) $= \frac{10}{3}$ instead of $= \frac{8}{3}$, causing thereby a small error in the numerical answer, which should ⟨be⟩ 1.809 instead of 2.0942 = the expression $c\sqrt{\frac{2h}{g}}$, making consequently the true distance 14461 Feet. Notwithstanding my utmost care in the ⟨actual⟩ operation of my solutions errata have crept in unawares which I hope you will excuse.

Perhaps the following new question may merit your notice. I cannot send you a solution with it, having a multiplicity of Business on my hands at this present.[1]

...

As I find the school here not calculated[a] ⟨to answer my original prospect viz⟩ the reception of Children of respectable People[b], on account of their unavoidably mixing with ⟨and consequently catching the habits of⟩ inferior Boys, and, which is still more objectionable[c] those on the foundation, I purpose[d] therefore to remove ⟨at Michaelmas to⟩ the parsonage-house at Little Bedwin about 3 miles from hence, which Parish I shall supply with my own Church of Froxfield together with the Chaplainship of the somerset Hospital to which I have been lately elected by the Noblemen & Gentlemen its Trustees.

Tho' this circumstance of removal change[e] my view of things here, yet I shall continue to keep alive the hope of one day or other becoming a near neighbour to my[f].[2] With the truest sense of the many Civilities and kindnesses with which your polite attention to me, from time to time, were so particularly distinguished and with sincere regard & respect I am

Dear Sir

your most obliged

and very humble servant

L. Evans

I beg my respectful Comp[ts]. to the Ladies

---

[a] *deleted*: for.

[b] *altered from (first)* Genteel People; *(second)* respectable ?families.

[c] *deleted*: with.

[d] *altered from*: mean.

[e] *altered from*: changes.

[f] *deleted*: good Friend D[r]. Hutton.

[*The following is bracketed and marked in red pencil for deletion:*]

[Pray do you know of any Clergyman who[a], you think, would like this situation & enter into the spirit of it? If his Terms ⟨would⟩ exceed no more than 14 Guineas, ⟨to⟩ w[ch] I could not think of acceding[b], I will venter to say, he would[c] soon have a very flourishing school. For this neighbourhood abound[s] in Wealthy Farmers, Malsters, Mealmen &c whose sons would be his only[d] Boarders for the reason already given.]

Notes

For Lewis Evans see note to Letter 17.

[1]An editor's note in the *Ladies' Diary*, repeated year after year, requested that all problems proposed should be accompanied by solutions. Apparently Evans believed an exception might be made; compare Letter 23, in which he provided no specimen of the kind of solution he believed the problem deserved.

[2]The phrase 'near neighbour', together with the evidence in Letter 40 that he had visited the Royal Observatory, perhaps indicates that Evans hoped to become an assistant at that institution.

Letter 42. LEWIS EVANS to CHARLES HUTTON, 29 September 1788

Oxford, Museum of the History of Science, MS Evans 31: draft.

Hungerford Sep[r]. 29th. 1788

Dear Sir

I was exceedingly astonished to read in a Letter I received a few days since from M[r]. Crocker that M[r]. Francis of Hook near Odiham[1] informed you that "I have railed against your Arithmetic[2] in public Company &c". Being conscious of my own Innocence & therefore to extinguish such an Idea from your Mind, I have requested the favour of all those who were present during the time I was in public Company with him to meet me this day at the

[a] *altered from*: whom.

[b] *deleted*: to.

[c] *altered from*: will.

[d] *altered from*: chiefly his.

Town Hall, that I might lay before them Mr. Francis's accusation. I cannot describe to you the detestable & contemptuous Light in which they looked upon Mr. Francis, for having in so wantonly false & scandalous a manner traduced my Character; which led them spontaneously to draw up the inclosed declaration,[3] which when read, I entertain no doubt, will honourably acquit me at the Bar of so liberal a minded Man as Dr. Hutton. That Mr. Francis & I had conversation about your Arithmetic & your other works I cannot deny, which was to this Effect, & judge you whether the most distant disparagement was given. He asked me whose Arithmetic I made use of? I told him Dr. Hutton's. He said the first impressions were very erroneous. I told him there were a few Errata in the 7th. Ed. some of which I corrected & communicated to you, together with a rule for perform[in]g Practice in a general manner. And that I meant as soon as the Hurry of removal was over, to point out some improvements in a few of the Rules.[4] I hinted to him where. He seemed to approve of it. After this he reprobated the common mode of Instruction by the present arithmetic publications & strongly recommended a Book of his, which, he told me, he had found by long Experience & Practice to be far superior to any Treatise of Arithmetic yet offered to the Public. I replied that I found no inconvenience from Dr. Hutton's Book, as no Boy made use of a Copy Book, but went through a Rule over & over till he was Master of it; and that I obliged my Boys to get each Rule by heart, by which means a ground work was laid for a future superstructure.[5] He rather sneered at my Plan. Before the Trustees were convened he & another Gentleman called to see the school & Premises, when upon your Name being mentioned in consequence of your Diary's lying on my Table, he said, he was intimately acquainted with you, had been at your House for weeks together & that you always sent him your Manuscripts to revise before you delivered ⟨them⟩ to the press. A singular circumstance attended his application for the school. He happened accidentally to meet one of the Trust at Newbury, who told him that I was going away from Hungerford school[6] & asked him if he knew of any Clergyman (or Layman who understood the Classics) who would like to succeed me? He answered that he knew of a Clergyman who would be very glad of the situation & was perfectly qualified for the undertaking, & that was a late Assistant of his.[7] Upon his return home he wrote to this Clergyman signifying the opening & recommending him in the warmest Terms to the Trustees. The very next day after he wrote this Letter, he came & applied himself, & said ⟨it⟩ was true he had recommended his late assistant but he knew he would not do, as he attended too much to pleasure to be a Schoolmaster. He spoke also very disrespectfully of his moral Character. Finding from this conduct he had ⟨but⟩ little conciliated Interest with the Trustees he withdrew his application by saying that something better had offered itself to his acceptance; which we found afterwards from his late assistant upon his application was an egregious falsehood. I mention this as one trait of his Character.

In the course of our conversation at the meeting I asked him whether he had dipped into astronomical matters? He replied he had. Being desirous of knowing a good method of using Hadley's Quadrant by Land I asked him how he made use of that Instrument?[8] Never did I see a Man so much confounded; he hummed something & shifted an answer. Those who heard, smiled, but I did not urge the matter further.

To this circumstance & the indifference I paid to the overweening opinion he had of himself I attribute the cause of his having so diabolically endeavoured to depreciate me in your good graces & Esteem, finding I corresponded with you. After such singular marks of your Friendship, so many obligations with which I have been favoured, I must be a Savage indeed to be guilty of the baseness and ingratitude to vilify you, Sir, or any of your works. So far from this that I have made it my Business in every School or Academy in this Neighbourhood and in others where I have had connexions of recommending them in the most jealous Terms[a].

I could mention other things that no ways redound to Francis[b] Reputation, but I forebear tiring your patience, by this time worn out, with a recital of the actions of so mean & contemptible, atrocious & vilainous a Fellow. I am D^r^. Sir with great respect

Your ever obliged and devoted humble ser^t^.

L. Evans.

P.S.

After so many complete Books on the subject, you would be surprised to read a Paragraph in the last Reading Paper among the Reading news, which I shall transcribe.

M^r^. Francis, Author of the Ciphering Book on a new Plan &c, intends soon to publish by subscr^n^. a complete Treatise on mensuration, Land Surveying Gauging, Navigation & Mechanics in two Volumes in Octavo

[a] *altered from*: manner.
[b] *altered from*: his.

price 8s.[9] Those who wish to become subscribers, are requested to address a line to M^{r}. Francis master of the Boarding School at Hook near Odiham where young Gentlemen are genteely accommodated & taught Penmanship, English Grammar, Latin, Greek, French, Drawing, Dancing, Arithmetic, Book-keeping & every Branch of the Mathematics on the most reasonable Terms.

Notes

For Lewis Evans see note to Letter 17.

[1]Mr Francis of Hook is identified in Wallis and Wallis, *Index*, 51 as William Francis (b. 1760), a schoolmaster based successively at Shinfield, Hook, and Basingstoke; he contributed to mathematical periodicals and subscribed to Lorgna, *Dissertation*; he was still active in 1796.

I have not identified Mr Crocker; Wallis and Wallis, *Index*, 34, lists several men of that name, none of whom can be identified with this man with any confidence. He may possibly have been the teacher and surveyor Abraham Crocker who in 1806 published a successful *Elements of Land Surveying*, expressing in it admiration for Hutton's works: MPHE.

[2]Hutton, *Guide*.

[3]See Letter 43, the draft in Evans's hand of this supposedly spontaneous declaration.

[4]The seventh edition of Hutton's *Guide* was dated 1785. Although Evans had mentioned in Letter 40 the possibility of conveying to him 'emendations of some of the general Rules', no letter from him is extant containing the specific corrections and rule mentioned here, and the 1788 eighth edition does not seem to contain anything new relating to the rule of practice. (The rule of practice relates to arithmetical problems where the student is asked to find the cost of a given quantity of goods, the cost of one unit (one yard, one pound, one bushel) being given: or vice versa.)

[5]This would have been an atypical classroom practice. Copy books or 'cyphering books' survive in such numbers as to suggest that their use was widespread if not ubiquitous, and descriptions such as Walkingame, *Tutor's assistant*, A2^{v}, and Hutton, *Guide*, ii–iii, confirm this impression. See Denniss, *Figuring It Out* and Ellerton and Clements, *History of the Royal Mathematical School*; also on the (comparable) American context Ellerton and Clements, *Rewriting* and Ellerton and Clements, *Extraordinary Cyphering Books*.

[6]See Letter 41.

[7]In Letter 43 this assistant was named as a Mr Dawson, whom I cannot identify. (John Dawson (1735–1820: ODNB) was still practising as a surgeon in Sedbergh at this time, and of the several other Dawsons listed in Wallis and Wallis, *Index*, 37–8, none can be readily connected with Francis.)

[8]Hadley's so-called quadrant—strictly an octant—was an instrument broadly similar to the navigator's sextant and used primarily for finding the altitudes of distant and/or astronomical objects.

[9]I have not identified either of these publications, and suspect that the proposed subscription volume never appeared.

Letter 43. The Trustees of the Grammar School at HUNGERFORD to CHARLES HUTTON, 29 September 1788

Oxford, Museum of the History of Science, MS Evans 31: draft in Lewis Evans' hand.

To Charles Hutton Doctor of Laws, Fellow of the Royal society and Professor of Mathematics in the Royal military Academy, Woolwich.

We the underwritten, the Trustees of the Grammar School at Hungerford have been given to understand that a Mr. Francis of Odiham[1] ⟨in the County of hants⟩ has informed you, Sir, that the Revd. Mr. Evans has "railed against your Treatise of Arithmetic[2] in public Company". In justice to Mr. Evans & to[a] clear him from such a malevolent imputation we think it a duty incumbent upon us to declare, that upon Mr. Francis's coming here to offer[b] himself a Candidate for[c] the School, a Meeting of the Trustees was holden to receive Mr. Francis's recommendations &c when Mr. Evans attended. Mr. Evans & Mr. Francis, which was the only time they were in public ⟨Company⟩ had some conversation on different subjects, but nothing was mentioned by Mr. Evans either disrespectful of you, Sir, or your Arithmetic. And ⟨we⟩ believe that Mr. Francis's accusation is grounded in licentious falsehood.[d,e,f] as[g] we beg leave further to declare that so far from Mr. Evans's railing against your Treatise, it is the very system of Arithmetic he has constantly used; of which & other Works published by you, several of us have often heard Mr. Evans speak in the highest Terms of approbation[h] ⟨and praise⟩. In Testimony of this declaration we have set our Hands this 29th. day of Sepr. 1788.

Notes

This letter relates to the incident described more fully in Letter 42; for Lewis Evans see note to Letter 17.

[1]For Mr Francis see note to Letter 42.

[2]Hutton, *Guide*.

---

[a]Deletion obliterated.

[b]*altered from*: offering.

[c]*altered from*: tow[ards].

[d]*altered from*: in falsehood.

[e]*deleted*: and from.

[f]*deleted*: & do appear to ? as the Revd Mr ?Dawson a late assistant of his & who he had treated ill, was appointed Master.

[g]*altered from*: And.

[h]*altered from*: well.

Letter 44.
LEWIS EVANS to CHARLES HUTTON, 9 December 1788

Oxford, Museum of the History of Science, MS Evans 31: draft.

Decem^{r}. 9th. 1788

Dear Sir

My[a] confusion & long unsettled state after removal hither[1] has been the cause of my not having acknowledged the receipt of your last very kind favour sooner. While it afforded me great pleasure ⟨& satisfaction⟩ of having totally obliterated the very unfavourable impression which that incidious Fellow, Francis made upon your mind so prejudicially against me, indeed the consciousness of my Innocence & rectitude of Conduct anticipated that satisfaction, under the Idea of my justification, it gave me great concern to hear of your confinement by so severe attack of Illness. I sincerely hope by this time you are perfectly restored to your former good health. I should conceive, however, a Transition from the Atmosphere of Wolwich for a few weeks might be of essential Service to the permanent establishment of your Health.[2] Shall I then request the favour & pleasure of your Company to spend what time you can of the approaching Vacation at my Cottage at Little Bedwin? I shall contribute my best endeavours to make your excursion as agreeable as I am able. You will come to a Bachelor's House, as it were, for I have buried poor M^{rs}. Evans in August last, of which interesting Event I may have informed you. I hope M^{rs}. Hutton & the Young Ladies are well,[3] to whom I beg my kind remembrance. and believe me to be D^{r}. Sir
your sincere Friend
and very obliged humble Ser^{t}.

Pray did you receive a Hare some time since, which I sent at the time of removing?

Notes

For Lewis Evans see note to Letter 17.

[1]That is, to Little Bedwin: see Letter 41.

[2]Subsequent letters from Evans do not mention that Hutton ever in fact visited him; Hutton did, however, escape the atmosphere of Woolwich Arsenal—of which he

[a] *altered from*: The.

complained bitterly in his edition of Simpson, *Select Exercises*, xxii—by moving first, in 1786, to Shooter's Hill and later to Woolwich Common.

[3]The implication is that Evans had become personally acquainted with Hutton's household, presumably on a visit to Woolwich. We do not know when this took place, but it was quite possibly during the same period as the visit to Greenwich mentioned in Letter 40 (dated in April 1788).

Letter 45.
LEWIS EVANS to CHARLES HUTTON, 8 April 1789

Oxford, Museum of the History of Science, MS Evans 31: draft.

Little Bedwin
near Hungerford Berks
April 8 1789

Dear Sir

Having since my removal hither[1] been employed in various occupations I am enabled to send but a few Solutions this year, and such as they are I present as a free will offering to my favourite old Lady.[2] I am sorry I cannot now approach her with a more substantial Tribute of respect & veneration. I am, Dear Sir,

with great truth
your most obliged
humble servant
L.E

P.S. I beg my most respectful Comp^{ts}. to the Ladies. Pray did you receive two Pheasants accompanied with a Letter from me in the Christmas Vacation.

...

Notes

For Lewis Evans see note to Letter 17.

[1]Evans moved to Little Bedwin about Michaelmas (29 September) 1788: see Letter 41.

[2]That is, the *Ladies' Diary*, addressed by many contributors as 'Lady Di'.

Letter 46.
LEWIS EVANS to CHARLES HUTTON, 26 May 1789

Oxford, Museum of the History of Science, MS Evans 31: draft.

Little Bedwin
near Hungerford
May the 26th. 1789

Dear Sir

A few weeks since I received from a M^r. Taylor proposals for publishing by subscription a Table of Logarithmic sines & tangents to every second of a degree, taken at sight.[1] To which he prefixes a Table of Logarithms from 1 to 100000. He says in his paper that the undertaking has been announced to the public in the Journal des Scavans for June 1786. To all this I make no doubt you are no stranger. But it appears to me rather extraordinary that M^r. Taylor should prefer a foreign print for announcing his performance. I should have known nothing of his intended publication, had he not sent me his printed proposals, addressed to me in a very curious & singular way without any apology for the freedom he took with a total stranger. That I may excite your smile I shall here transcribe his address. "Rev^d. M^r. L. Evans Mathematician of Hungerford Wilts. single sheet". within were only these words "From the author". I am obliged to him for the compliment he pays me of being mathematician of Hungerford, taking it for granted perhaps that solving a few of the easier questions in the Ladies' Diary constitutes a person mathematician permanent of the place in which he lives. I shall esteem myself much obliged to you for informing me what merit this intended publication lays claim to. As I have some months ago purchased your Tables[2] I would not wish to go to the expence of having an inaccurate transcript of part of them. I must own the having of sines & tangents accurately computed, which is here a doubt,[3] to seconds of a degree exhibited at Sight is very commodious. But why should M^r. Taylor, with all proper deference to his abilities, put us to the unnecessary charge of unavoidably buying a Table of Logarithms when we already possess in a state of perfection, at least as far as those Numbers are carried? Had he presented to the public a Table of Logarithms to 10 or 12 places of the respective Numbers he had done us additional service; which would furnish him with a good plea for prefixing such a work to his new Tables of sines & Tangents. Indeed a Table of this sort would be very useful in the reducing many algebraic expressions to Numbers, as you must be sensible, and also in the extraction of Roots.

I anticipate much pleasure in the perusal of your Book of trigonometrical Tables of the Lengths of Arcs in the aliquot parts of the Radius.[4] Pray, favour me with an early account of their completion.

I think I expressed to you in a former letter my wishes of being acquainted with the construction of what is called the gridiron pendulum. There is nothing satisfactory in the Cyclopædia.[5] Can you refer me to any work descriptive of it. I am the more earnest as I feel ambition in having an accurate chronometer. If you can give me some hints by favouring me with a rough Draught, it may be sufficient.

With the most sincere[a] respect I am, Dear Sir.
Your[b] obliged
humble Servant
L. Evans.

I beg my Compliments to the Ladies.

Notes

For Lewis Evans see note to Letter 17.

[1]Taylor, *Table*. Michael Taylor (dates unknown) was also the author of a sexagesimal table published in 1780 by order of the Commissioners for Longitude: MPHE. The logarithmic table was subsequently advertised in the *Nautical Almanac* for 1793.

[2]Hutton, *Tables*, which contained among other matter logarithms from 1 to 100,000, natural and logarithmic sines, tangents, secants, and versed sines for every minute of a degree, and sines and tangents for every second of a degree.

[3]Evans' hand is not particularly unclear at this point, but his meaning is obscure. Perhaps 'here in doubt' is the most plausible emendation.

[4]Evans referred to the proposal mentioned in the introduction to Hutton, *Tables* (see Letter 18), where he had presumably seen it.

[5]Chambers, *Cyclopædia* (1728), of which new editions continued to appear during this period. The 'gridiron' pendulum was a temperature-compensating pendulum consisting of a grid of steel and brass rods, invented by the celebrated horologist John Harrison (1693–1776).

[a]Deletion obliterated.

[b]*deleted*: ?very.

Letter 47.
LEWIS EVANS to CHARLES HUTTON, 27 August 1789

Oxford, Museum of the History of Science, MS Evans 31: draft.

Little Bedwin
Augst. 27th. 1789.

Dear Sir

I am commissioned by a Gentleman to enquire for a situation for a young Lad, an orphan, to be qualified for the naval Life; as you may most probably know of a seminary in that way I shall be much obliged to you for informing me of one which you can recommend. The questions to the Master are.

1. What are his Terms for Board & Tuition?

2. Whether his School is confined to Navigation alone? As French & dancing will be required. The Lad is not to learn the Classics.

3. Whether Boys are taken care of during the Holidays, if required & what is the additional allowance? This is a most material point, as the Lad is to be wholly kept at School, his Guardian having no convenience for his accommodation during School vacss.

The young Gentleman's name is Davy, Son of the late Major Davy, Interpreter to Governor Hastings.[1]

I am truly obliged to you for your last favour respecting M^{r}. Taylor,[2] which I should have ere now acknowledged. I have not heard from him since.

I have been lately on an excursion to Portsmouth & the Isle of Wight. The farthest place we visited, (for two Gentlemen were with me) on that Island was Causbrook Castle near Newport. where Charles the first was kept Prisoner for about a twelve Month. It is situated upon an Eminence commanding an extensive prospect towards Hampshire. The fortification at Portsmouth, the Dock Yard & the grand magazine of Stores &c for the destruction of the human species afford a sight truly wonderful. The Academy in the Dock yard[3] is not a magnificent but decent pile of Building. Having no acquaintance, we did not go within it. From this place we took a wherry & sailed round the Harbour, where there might be no less than 40 ships lying in ordinary, as the sailors term it. We went on board the Barfleur, the port Admiral's Ship; every part of which was shewn us with remarkable alacrity & attention. Another astonishing effort this of human genius. To be brief, I have not been better pleased with an excursion scheme, almost every object appearing interesting, for we took Winchester, South-hampton & Salisbury in our Route. The Cathedral of the Latter place is now undergoing a thorough repair & beautifying under the direction of our Diocesan D^{r}. Barrington Brother to Lord Barrington.[4]

With most sincere regard & respect
I am, Dear Sir,
Your most obliged
humble servant
L. Evans

I beg my respectful Compliments to Mrs Hutton & the Young Ladies.

Notes

For Lewis Evans see note to Letter 17.

[1]Major William Davy (d. 1784) of the Bengal army was Persian secretary to Warren Hastings (1732–1818), Governor-General of Bengal. His son, William Gabriel Davy (1780–1856), in fact went on to a distinguished career in the army after education at Eton.

[2]For Mr Taylor see note to Letter 46.

[3]The Royal Naval Academy, established at Portsmouth Dockyard in 1733.

[4]Shute Barrington (1734–1826) was at this time bishop of Salisbury; his brother was the politician William Wildman, second Viscount Barrington (1717–93): ODNB.

Letter 48.
LEWIS EVANS to CHARLES HUTTON, 12 May 1790

Oxford, Museum of the History of Science, MS Evans 31: draft.

May 12th. 1790

Dear Sir

I now send you my mite for the Treasury of Old Lady Di.[1] I wish it had been in my power to transmit more, but various avocations have precluded that pleasure. I have, as it is, to request your excuse in being so late with my packet.[2] I am, Dear Sir,

Your most obliged
and most devoted
humble servant
L. Evans

...

Notes

For Lewis Evans see note to Letter 17.

[1]That is, the *Ladies' Diary*.

[2]It was stated in edition after edition of the *Ladies' Diary* that contributions received after 1 May could not be used. Evidently Evans believed an exception could be made for him; compare Letter 41 and the note there.

Letter 49. CHARLES HENRY VIGNOLES to CHARLES HUTTON, 31 October 1790

Portsmouth, History Centre, Vignoles Papers, 1072A/App. 1: ALS (our source). Also printed in Vignoles, *Infant Ensign* as Appendix 1.

Greenwich 31st Octr. 1790

Sir

The subject I am about to Address you on is of so interesting a Nature I [*sic*] that I found I could not explain myself Personally. I therefore take this Method of applying to you for your Approbation of a Circumstanc[e] on which the happiness of my Life depends, but as that is in my Mind a secondary consideration[a] when compared with that of your Lovely Daughter I will acquaint you with my Situation & circumstances[.] In the manner of the Character I hope Always to maintain that of a man of honour & Integrity, When this is done you will be a Judge knowing the World whether I shall be a proper Guardian for Life for your Daughter[.] I am too well Acquainted with the World not to know that a competency is Absolutely necessary to the the [*sic*] happiness of persons who have lived in the sphere of Life I have been used to, but at the same time I do not Aspire to Affluence — I long for happiness and Camilla is Alone Capable of bestowing it. I now come to the explanation and only expect what I am sure to find the same Candour on your Part, as you have found on mine, concious of the Rectitude of my intentions I trust my cause in the hands of the Arbitrator of my Fate. By a chain of circumstances which I can easily explain, when I Purchased the Adjutancy I was deficient in the Purchase money almost the whole Sum which is £400 but I was so well esteemed in the Reg^{t}. that the money was

[a] *deleted*: in.

Advanced for me. As I was not permitted to sell when I Got my Company I was obliged to my Friends for the Money & which I am now indebted to them in my Present situation I pay a small Portion yearly & they are satisfy'd, But shou'd I change they wou'd naturally expect that every Obligation of that Nature shou'd be discharged.

Thus Sir am I situated in Point of Fortune & by this you may see it wou'd be the heighth of cruelty in me if I had it in my Power to involve a lovely young Creature in distress of which I have seen too many instances if she has not wherewithal to make our Situation in Life easy.[1] — I cannot marry without Fortune & I never will for it. My choice is fixed & if disappointed now I shall drop All thoughts of it[.] Indispensable business prevented my writing Yesterday as I intended. I anxiously wait you[r] Answer as on it my Fate depends — whatever it may be I shall always remain with the Greatest Respect & Regard your most Obed^{t}. very Humble Servt:

C.H. Vignoles

Note

Charles Henry Vignoles (d. 1794) was an officer in the 43rd Regiment of Foot, said to be descended from a Huguenot family. He married Charles Hutton's daughter Camilla (for whom see note to Letter 16) later in 1790 and the couple moved to Ireland before going to the West Indies in the deployment of 1793; both died in Guadeloupe in 1794. See Letters 61, 66, 78, 93, and 110.

[1]Evidently Hutton took this very broad hint and provided Camilla with the 'wherewithal' Vignoles' situation and lifestyle required: the couple were married on 18 December 1790, at St Mary Magdalene, Woolwich.

Letter 50.
LEWIS EVANS to CHARLES HUTTON, 28 April 1791

Oxford, Museum of the History of Science, MS Evans 31: draft.

April 28th 1791

Dear Sir

I have sent you a few solutions for the Ladies' Diary and remain, Dear sir, your obliged humble servant

L. Evans

...

some time ago I sent you a new question respecting an Elliptical picture frame, which, if you approve of ⟨it⟩, I shall be obliged to you for inserting.

Upon reconsidering[a] ⟨Quest. 917⟩ I would wish to correct my solution by saying something similar to this;

...

should you think the following easy question worthy of a place in either the Diary or Supplement for the exercise of young proficients, I shall at my first Leisure send you its solution.

...

Note

For Lewis Evans see note to Letter 17.

Letter 51.
LEWIS EVANS to CHARLES HUTTON, 7 May 1791

Oxford, Museum of the History of Science, MS Evans 31: draft.

May 7th. 1791

Dear Sir

I take the opportunity of transmitting to you a solution to my last new question and also another new question with its solution, which I conceive to be[b] of real use, I am most ⟨obligedly &⟩ respectfully
yours L. Evans.

...

Note

For Lewis Evans see note to Letter 17.

[a] *altered from*: reconsideration of.
[b] *altered from*: ?as.

Letter 52.
CHARLES HUTTON to WILLIAM FREND, 21 May 1791

Cambridge, University Library, MS Add. 7886/117: ALS.

Woolwich, May 21, 1791.

Dear Sir,

I find you do not yet fully take the idea I threw out concerning Des Cartes's method of dissolving a biquadratic equation into two quadratics; & I am much surprised to observe that Wallis says he does not know how Des Cartes came by his rule;[1] for I don't think any thing can be more simple & natural, or easy, than that wch I hinted, &[a] wch I am pretty sure was the method used by that author, viz, to find out the <u>form</u> & the <u>coefficients</u> of the two quadratic equations, whose[b] roots should be the same as those of the proposed biquadratic. This method is simply no more than this, viz, to generate ⟨or produce⟩ a new equation, of the same height with any proposed one, by multiplying together two or more ⟨assumed⟩ equations of lower orders, wch equations have any general assumed coefficients; then by comparing or equating the corresponding coefficients of the proposed equation, with those of the produced equation,[c] the one of these two equations will thereby be made equal to the other in all respects, the terms & their signs each to each, & consequently their roots also; and also the said comparison of the coefficients will determine the values of the assumed coef^s. of the several assumed lower equations. And this method is general for all equations whatever, or whatever their height & number of terms & signs of those terms may be. Nor is ⟨there⟩ any necessity to make ⟨or prove⟩ any such suppositions as in the 7th article of your paper, or any other whatever; for the equations arising from the comparison or equating of the coef^s. of the proposed & produced equations, are all independent of one another, & the n^o of them is always equal to the n^o of the unknown or assumed coeff^s. of the assumed & lower equations

Now Des Cartes was ⟨fully⟩ possessed of this general method, and could have applied it to other cases & heights of equations besides that single case wch he has given in his book, viz of dissolving a biquadratic equation into

---

[a] *deleted*: by.
[b] *deleted*: product should be equal.
[c] *deleted*: will.

two quadratics, & that whether his equations wanted the 2d term or any other terms, or not: in the example he has given, he[a] supposed the 2d term absent, or taken away, ⟨only⟩ because he found that supposition made the rule much easier & simpler than it would otherwise be; as you may find on trial.

But now to apply this method to the case in hand, viz to divide the proposed biquadratic $x^4 \pm qx^2 + rx + s = 0$[b] into two quadratic factors. It is evident that the first term of each factor may be x^2, because their product is to be x^4, & theref. the form of those two factors will be

$$\begin{cases} \quad x^2 + yx + z = 0, \\ \& \ x^2 + vx + w = 0; \end{cases}$$

that is, as to the forms of the terms, independent of the signs of them + or −, wch are circumstances that will be discovered afterwards by the comparison or aquating of the coefficients, as above mentioned.

Again, to discover,[c] in the same way, what are to be ⟨the⟩ values & the signs of the coef^s. y & v of the 2d terms of these two quadratics, it is only to be considered that when these two quad^rs are multiplied together, then the 2d term of the product is $\overline{y+v}.x^3$, & because the corresponding term of the proposed biquad. is absent or nothing, theref. this also is to be $= 0$, i.e. $\overline{y+v}.x^3 = 0$, or $y+v = 0$, or $v = -y$; i.e. the coef^s. of the said two 2d terms yx & vx of the assumed quadratics, must be alike, or the same quantity, ⟨but different signs,⟩[d] the one affected with the sign +, & the other with the sign −; & hence then the form of the said two equa^s. will be

$$x^2 + yx + z = 0,$$
$$\text{and } x^2 - yx + w = 0.$$

Thirdly, to find the values & signs of the 3d or last terms z & w of these assumed quadratics, multiply the two equa^s. together, & compare the last ⟨three[e]⟩ terms of the product with the last ⟨three[f]⟩ of the proposed[g] biquadratic: thus, the product of these two quadratics is $x^4 \pm \overline{z + w - y^2}.x^2 + y.\overline{w - z}.x + wz = 0$, and the proposed equa^n. is ... is [*sic*] $x^4 \pm qx^2 + rx + s = 0$.

[a] *deleted*: only.
[b] Hutton used the notation ∗+ for ±.
[c] *deleted*: what.
[d] *deleted*: only.
[e] *altered from*: two.
[f] *altered from*: two.
[g] *altered from*: assumed.

Hence, comparing or equating the last three terms to each other, we have these three independent equations, viz.

$$\begin{cases} w+z-y^2 = q, \\ \overline{w-z}.y \quad = r, \\ \& \ wz \quad\quad = s. \end{cases}$$

And by resolving or reducing these three equa[s]. according to the common rules of Algebra, there will be had the values of w, z, & y, in terms of q, r, s, & also the relation of their signs, the same as stated by Des Cartes. Thus,[a] by transposing y^2 in the 1st equ[n]. & dividing by y in the 2d; give these two

$$\begin{cases} \quad w+z = q+y^2, \\ \& \ w-z = \frac{r}{y}: \end{cases}$$

then by adding these two together, & subtracting the one from the other, & dividing the results by 2, we have

$$\begin{cases} w = \frac{1}{2}q + \frac{1}{2}y^2 + \frac{r}{2y}, \\ z = \frac{1}{2}q + \frac{1}{2}y^2 - \frac{r}{2y}; \end{cases}$$

w[ch] are the values of those two quantities w & z, ⟨& of the signs,⟩ as stated by Des Cartes.

Finally, to find the value of y, from the 3d equa[n]. $wz = s$, substitute the above values of w & z here instead of them, and it will be

$$\frac{q+y^2+\frac{r}{y}}{2} \times \frac{q+y^2-\frac{r}{y}}{2} = s,$$

or

$$\overline{q+y^2}|^2 - \overline{\frac{r}{y}}|^2 = 4s,$$

or

$$q^2 + 2qy^2 + y^4 - \frac{r^2}{y^2} = 4s,$$

or

$$q^2y^2 + 2qy^4 + y^6 - r^2 = 4sy^2,$$

[a] *deleted*: the.

or

$$y^6 + 2qy^4 + \left.\begin{matrix} q^2 \\ -4s \end{matrix}\right\} y^2 = r^2;$$

w[ch] is the cubic equation as given by Des Cartes, & w[ch] comprehends all the possible cases or varieties of the signs of the terms of any proposed biquadratic equation, making $2q$ here have the same sign as q there (whether + or −), but $4s$ here the contrary sign to s there: as to the signs of q^2 and r^2, they are always +, whether the signs of q & r be + or −, because the square of either is +.

And this is the complete investigation of that rule ⟨at once⟩ for all the cases of the signs of biquadratic equations. You may amuse yourself with any other ways of doing the same thing, or with any varieties & changes of signs and forms; but the above is Des Cartes's, & very general and short it is.

I am, Dr Sir, You[r m]ost obed[t] hble Serv[t].

Cha. Hutton.

P.S.

I am desirous to ask you a question in Law affairs, w[ch] doubtless you can well resolve me.

A Court Baron & Court Leet was held here lately on the affairs of the Common Grounds & other matters, at w[ch] one of the Jurors swore that, in building a certain wall, I had inclosed an adjoining Road,[2] in doing w[ch] he has made a false oath, as the contrary is notorious to the builders of the wall, & to every other person, who can prove it. Now this piece of perjury is so base, that I wish to know whether & how the Juror may be convicted of it, how probably punished, what may be the expence of such a prosecution, & who must bear that expence, &c, &c.

C.H.

I hope to attend the Club[3] on Friday next.

Notes

William Frend (1757–1841), religious writer and actuary, pursued an active interest in mathematics alongside his other activities, which included political controversy, popular writing on astronomy, and mathematics tutoring (he taught among others Annabella Milbanke, the daughter of Judith Noel, for whom see note to Letter 94): ODNB.

[1]The technique described appeared in Descartes, *Géométrie*, 385–7; Wallis, *Algebra*, 208, gave the rule and stated (in the context of Wallis's customary depreciation of Descartes) that 'How he came by that Rule, he doth no where tell us'. Frend did not discuss this particular subject in his 1796 *Algebra*.

[2]Hutton purchased a parcel of land by Woolwich Common in 1789 or 1790 and erected about eighteen houses over the following fifteen years; this passage hints that some degree of controversy attended the new buildings, but there seems to be no evidence that the matter went any further. Hutton, 'Note on the Divining Rod', 466; Guillery and Saint, *Woolwich*, 3–6; Bruce, *Memoir*, 21.
[3]For the 'Friday' Club see note to Letter 20.

Letter 53.
Lewis Evans to Charles Hutton, 23 May 1791

Oxford, Museum of the History of Science, MS Evans 31: draft.

May 23^{d}. 1791

Dear Sir

Having received the Diary Supplement but this day ⟨by an untoward circumstance⟩ I could not[a] transmit you the following solutions (if they may be termed such) otherwise than ⟨in the state⟩ you see them, without a much longer protraction of Time.[1] I am, Sir,
your most obliged
humble servant
L. Evans

...

You probably may know that a M^{r}. Robertson of Christ Church Oxford is about to publish a ⟨Latin⟩ Treatise of Conic Sections in one volume quarto. This is the Gentleman who is sub-lecturer[b] for D^{r}. Smith Professor of Mathematics at Oxford.[2] As far as I can learn M^{r}. R. deduces the properties of the Conic sections from the Cone alone, which in the judgement of some is most natural.

Notes

For Lewis Evans see note to Letter 17.
[1]Once again Evans implicitly asked for an exception to the usual deadline for material for the *Ladies' Diary*: see note to Letter 48.

---

[a] *deleted*: ?have.
[b] *altered from*: reads.

[2]John Smith (1721–96) was the Savilian Professor of Geometry at Oxford at this time. Abram Robertson (1751–1826), mathematician and astronomer, did indeed deputize for his patron Smith, and succeeded him as Savilian Professor in 1797; in 1810 he became Savilian Professor of Astronomy and Radcliffe observer. The work alluded to was Robertson, *Sectionum conicarum*. ODNB; see also Letters 62 and 74.

Letter 54.
Lewis Evans to Charles Hutton, 12 July 1791

Oxford, Museum of the History of Science, MS Evans 31: draft.

July 12th. 1791

Dear Sir

I have had a call from home for a few days, or I should have acknowledged the receipt of your favour before.

[a]I am exceedingly obliged by[b] your kind invitation to Woolwich, but a variety of[c] circumstances totally preclude the pleasure of accepting[d] at present.

Your having written[e] a Treatise of Conic Sections[1] was an Intelligence of which, before, I was not apprised, as I ⟨hardly⟩ never see any other paper besides[f] the London Chronicle,[g] ⟨in which it has not been advertised⟩.[2] It must be confessed that the fewer the principles are of any science, with the more facility it becomes[h] understood. We have a good[i] Instance of this in Simpson's Geometry,[3] a system founded upon as few Axions, postulates &c as it might be[j] possible the art of man could devise. The properties of the Conic Sections being[k] ⟨?Initially⟩ built upon speculative Geometry, an Introduction to the attainment of those properties cannot be sufficiently

[a] *deleted*: your h. . . .
[b] *altered from*: to you for.
[c] *deleted*: interesting.
[d] *altered from*: that pleasure.
[e] *altered from*: published.
[f] *altered from*: but.
[g] *deleted*: which I take three times a week. In this paper there are very few advertisements.
[h] *altered from*: its ?super ?structure is.
[i] *altered from*: sure.
[j] *altered from*: was.
[k] *altered from*: Conic Sections being.

valued, when it[a] removes every difficulty that has[b] hitherto perplexed & even given disgust to almost every Beginner in this Branch of knowledge.[c4]

I certainly must join with you in opinin[g] that the deductions of the Sections from the cone ⟨itself⟩ is the most natural & rational.[5]

I remember when the Marquis D'L'Hospital's Treatise[6] was put in my hands, which was the first I[d] ever saw, I made an observation to my Tutor that I thought I was beginning where I should end. Why so, says he. How do I know – I rejoined – that the mode directed here of describing this curve, forms a[e] parabola & therefore[f] a conic section. The only answer I received was, that I must take it for granted that it was such a curve. If this be previously admitted[g]; to be sure, the properties of the several sections ⟨of the Cone⟩ are very ingeniously deduced and admirably demonstrated by several very able mathematicians. But this does not give us[h] a commanding view of the subject, in our[i] first entrance upon it. The[j] modes of demonstration ⟨too⟩ have been[k] a matter[l] of objection. Some prefering the analytic to any other. For my part I apprehend the neatest & concisest way must be something similar to what we find[m] adopted by the excellent M[r]. Simpson. Hamilton, Jack & D[r]. Simson are[n,o] prolix; tho' the latter is an elegant composition of the sort.[7]

I have been lately engaged in casting & polishing Specula for a Gregorian[p] Telescope, in which I have succeeded beyond my expectations.[q] In this mechanical pursuit I chiefly attended to ⟨the Rev[d]⟩ M[r]. Edwards's directions.[8] It is remarkable enough[r] that an elliptical tool, the Diameters

---

[a] *altered from*: if it.

[b] *altered from*: had.

[c] Two or three distinct deleted passages follow, largely obliterated but containing a legible reference to an (unspecified) error committed by 'Mr Vince of Cambridge' in the eighth proposition of his 'Elements of Conics'.

[d] *deleted*: th....

[e] *altered from*: is a.

[f] *altered from*: consequently.

[g] *altered from*: granted.

[h] *altered from*: me.

[i] *altered from*: ?this.

[j] *altered from*: Some.

[k] *altered from*: ?are.

[l] *deleted*: too.

[m] *deleted*: ⟨adopted⟩ in Simpson's.

[n] *deleted*: tedious &.

[o] *deleted*: very.

[p] *deleted*: reflecting.

[q] *deleted*: I have.

[r] *altered from*: very wonderful.

of which being in the proportion of 10 to 9, should form the speculum of a parabolic figure. Of this we were, I believe, promised a demonstration,[a] had Mr. E. lived.

With compliments of respect & esteem to Mrs. Hutton, the Ladies & yourself, I am Dear Sir

your greatly obliged servant

L.E.

P.S.

Mr. James Junr.[9] has given me a most melancholy account of a late catastrophe at Greenwich which happened to a Diary Correspondent, with which you may be acquainted thro' the medium of Dr. Maskelyne.[10] The young gentleman paid me a visit last Christmas seemed ?perfe[c]tly chereful & composed in mind.

Notes

For Lewis Evans see note to Letter 17.

[1]Hutton, *Conic Sections*.

[2]Hutton, *Conic Sections* seems to have been intended primarily for use at the Royal Military Academy at Woolwich; I have not located any advertisements for it in the public newspapers.

[3]Simpson, *Geometry*.

[4]The reference was evidently to Vince, *The Elements of the Conic Sections*, by the Cambridge-based astronomer and mathematician Samuel Vince (1749–1821), subsequently (from 1796) Plumian Professor of Astronomy and Experimental Philosophy.

[5]See Letter 53.

[6]l'Hôpital, *Conick sections*, in which the first definition (pp. 1–2), described how to draw part of a parabola using a rule, square, thread, and pin.

[7]Evans alluded to Hamilton, *De sectionibus conicis* (published in English as Hamilton, *Conic Sections*), Jack, *Conic Sections*, and Simson, *Sectionum conicarum* (published in part in English as Simson, *Conic Sections*).

[8]John Edwards (*c.* 1747–84) (see Croarken, 'Mary Edwards: Computing for a living in 18th-century England', 9–11) produced, under the patronage of the Board of Longitude, 'Directions' for the making of telescopes, printed as an annexe to the 1781 *Nautical Almanac* and at least twice as a separate pamphlet. The full title of the text was

> Directions for making the best composition for the metals of reflecting telescopes; and the method of casting, grinding, polishing, and giving them the true parabolic figure.

[9]Not identified.

[10]For Nevil Maskelyne see note to Letter 13. I have not been able to trace the 'catastrophe' at Greenwich, which sounds from Evans' description like a suicide among the staff at the Royal Observatory.

---

[a] *deleted*: but, poor Man, he is gone after the Manner of.

Letter 55.
CHARLES HUTTON to WILLIAM HERSCHEL, 15 January 1792

London, Royal Astronomical Society, William Herschel Correspondence 13.H.34: ALS.

Woolwich, Jan. 15, 1792

Dear Sir,

On examining Mr Dalby's paper in the Philos. Trans. 1791,[1] I am of opinion he has made it probable that the two diameters are nearly as he makes them at the bottom of pa. 239; that is, on the supposition that he has performed all his calculations accurately. But, after such mistakes in computations as have recently happened,[2] one does not know where to rest a sure dependence. From the table he has given in pa. 238, it seems that the measurements in Middle latitudes are more to be depended on that [*sic*] those at the equator & polar circle; &, it seems that the former measures give a less magnitude of the axes than the latter. From Mr Dalby's diam^{s}. reduced to miles, I deduce as follows: viz,

Earth's Axis	$7894\frac{2}{3}$ miles
Equatorial Diam.	$7931\frac{2}{3}$
Mean Diam.	$7913\frac{1}{6}$
Mean Circumf.	24860
Mean length of a Degree	$69\frac{1}{18}$

The ratio between his diam^{s}. in the nearest whole numbers, ⟨that differ by unity or 1,⟩ is that of 215 to 214. I am,

Dear Sir,

Your obed^{t} hble Servant,

Cha. Hutton.

Doctor Herschel
Slough
Windsor.

Notes

For William Herschel see note to Letter 21.

[1]Dalby, 'The Longitudes of Dunkirk and Paris'; the passage alluded to gave values for the polar and equatorial radii of the earth.

[2]I have not traced the reference of this allusion.

Letter 56. LEWIS EVANS to CHARLES HUTTON, 26 April 1792

Oxford, Museum of the History of Science, MS Evans 31: draft.

April 26th. 1792

Dear Sir

I have, at length, sent you my annual mite for the Treasury of the old mathematical Lady,[1] whose charms, with those of her chaste offspring, cannot but inspire and cherish a Love of Science.

I have the honour to be, Dear Sir,
your most obliged
humble servant
L. Evans

...

Note

For Lewis Evans see note to Letter 17.

[1]The *Ladies' Diary*; her 'offspring' being the *Diary Supplement* published annually from 1788.

Letter 57. LEWIS EVANS to CHARLES HUTTON, 1792

Oxford, Museum of the History of Science, MS Evans 31: incomplete draft.
Dated from the position of this letter in the bound volume of Evans's correspondence.

Dear Sir

I make no doubt but you have safely received my Packet of solutions for the diary in due time.[1] I have been since struck with a circumstance which I cannot clearly account for and which possibly may merit a place in the Diary. It is simply this.

It is asserted by M^r^. Simp[s]on in his Fluxions[2] that to determine the greatest area which can be contained under four given right-lines the figure may be inscribed in a circle. Again, it is suggested in the corollary to Theorem 11th. of his Maxima & Minima of Geometrical Quantities[3] that the same thing will be affected when ⟨the⟩ figure may be inscribed in a ⟨semi⟩circle of which the longest side will be the diameter. This is also implied in the solution to the 174 Question of the Ladies' Diary.[4] Now a trapezium of which the several sides are 10, 12, 15, and 20 will only answer the regimen of the former Theorem, that is, it may be inscribed in a circle, by Prob. 35 simpson's Select Exercises,[5] in which case its greatest area is had, but it cannot be inscribed in a semicircle of which its longest side is the diameter therefore its greatest area ⟨cannot⟩ in this case be had.[6] I must either egregiously misconceive the application of that corollary,[a] or the corollary itself as well as the solution is not generally true. I communicated this Geometrical embarrassment to the late M^r^. Cook[7] who was doub[t]less a good mathematician & he seemed much astonished at it yet could not at that moment develope the mystery. Should you judge this matter worthy of public investigation permit me to leave it to your superior Judgment as to the form in which it shall appear.[8]

[*The draft ends here at the foot of a page; a page or pages are probably missing from the volume of Evans' draft letters to Hutton.*]

Notes

For Lewis Evans see note to Letter 17.

[1]That is, Letter 56.

[2]Simpson, *Fluxions*, 39–42.

[3]Simpson, *Geometry*, 117.

[4]Question 174 appeared in the 1733 *Diary*: Hutton, *Diarian Miscellany*, vol. 2, pp. 22, 27–8.

[5]Simpson, *Select Exercises*, 133–5.

[6]In fact both Simpson, *Geometry*, 117, and a note at Hutton, *Diarian Miscellany*, vol. 2, p. 28, clearly state that a polygon, *one of whose sides is unknown*, has its maximum area when it is inscribed in a semicircle of which the diameter is the unknown side. Evans' example, a quadrilateral with four *given* sides, is not a counter-example to this.

[7]Not identified.

[8]I cannot discover that Hutton ever used this matter in the *Ladies' Diary* or the *Diary Supplement*.

---

[a] *altered from*: corollary.

Letter 58.
LEWIS EVANS to CHARLES HUTTON, 1793

Oxford, Museum of the History of Science, MS Evans 31: fragment of draft.
Dated from the position of this letter in the bound volume of Evans' drafts, and from the reference to Letter 54.

D[r]. Sir

I sent my packet of solutions[1] ⟨the latter end of April⟩ by a person who was going to London from this Neighbourhood, which I requested ⟨him⟩ to leave at stationers Hall[2] whence[a] I entertain no doubt of its[b] safe conveyance to your hands. Its contents were the result of as much Time ⟨as⟩ I could spare in that way. I think I told you about two years ago[c] of my success in casting & polishing mirrors for a Gregorian Telescope.[3] I have since fitted them in a Tube, which I found to be a work of more difficulty. As it was necessary the two specula should be ⟨situated⟩ perfectly parallel to one another & the small one to slide precisely in the axis of the greater and still to keep its parallelism to it[d]. I met also with some difficulties in the adjustment of the Glasses in the Snout. all which by[e] perseverance I overcame. The dimensions of the Telescope thus completed (excepting a stand) is 20 Inches focal distance, and aperture about 4. The magnifying power in its present state is considerable[f]; but by the application of other eye glasses, (as

[*The text breaks off at the foot of a page. A page or pages are evidently missing from the volume.*]

Notes

For Lewis Evans see note to Letter 17.

[1]The covering letter for Evans' solutions for the 1793 *Ladies' Diary* is not extant.

[2]A note in the *Ladies' Diary* requested that solutions and new problems be directed to the editor at Stationers' Hall, the seat of the Stationers' Company of London, which published the *Diary*. It seems probable that by this date Evans was sending his more personal letters directly to Hutton at Woolwich.

[3]See Letter 54.

---

[a] *altered from*: & which.

[b] *altered from*: their.

[c] *altered from*: since.

[d] *altered from*: ... axis of the Tube.

[e] *deleted*: indefatigable.

[f] *altered from*: 44.

Letter 59.
CHARLES HUTTON to ROWLAND BURDON, 18 June 1792

Plymouth and West Devon Record Office, 105/164: letter sent.
Appears to be in a scribal hand—possibly that of Isabella Hutton—except for the address and signature.

Woolwich
June 18, 1792

Sir,

Upon perusing M^{r}. Mackintosh's Section of the river, & the sections & inginious Observations of M^{r}. Richardson,[1] the following ideas & remarks have occurred

There are two main & leading articles of consideration, viz. 1st. the n^{o}. of arches, & 2^{d}. the situation or place of the bridge. — As to the first of these, or the n^{o}. of arches, it appears there are two opinions, some being for one large arch only, & others for 2 arches.[2]

The advantages in favor of one arch only are very considerable; as, first, it may be placed in any one situation that may be thought most convenient for the town or country, because it will leave the whole water-way free & open, for the passage of the keels & ships. 2dly, Its great height, of about 100 feet in the clear, will both allow the ships to sail easily under it, & it will raise the way over the bridge to a level with the banks on the north & south sides, which will be easiest for the carriages of any, as it will thus have neither ascents nor descents. And 3dly, It will be of the least materials & least expence of any, perhaps of about $\frac{2}{3}$ only of the expence of the bridge with 2 arches. Its disadvantage is the chance & fear of not succeeding in the construction of so large an arch, the difficulties & hazard increasing with the magnitude of it, especially if the architect be not very expert. —

As to the scheme for 2 arches, the advantages & disadvantages are just the reverse of those of one arch; viz, it will be attended with less hazard in the formation, but then it must have a pier in the middle of the river, will require a great deal of more labour & materials & expence; be also less convenient for the passage over it, by having a descent to[a] both sides; and be less convenient for the navigation below it, from the lowness of the arches, which circumstance may prevent you from using the best situation for the

[a]Source has *to from*.

town, viz, by occasioning it to be carried higher up the river to be above the births where the ships come to load. — Upon the whole consideration of these two circumstances, you will satisfy yourselves in the choice between[a] the two cases. Were it my own concern, I would chuse the former case, or that of one arch, as having several probable advantages over the other, especially if the arch be built of a proper form, so as to be an arch of equilibration,[3] and that the architect be sufficiently intelligent & skilful. Perhaps it might be adviseable to have some check over him, such as this, viz. that he shall have nothing for his trouble if the bridge should fall down after it is quite completed & the centres struck, or within a year after that time; but in consideration of its standing that trial, then his profits to be a little higher than usual. —

As to the situation; the best will be obvious to persons acquainted with the place & its Trade. In case of a single arch bridge, it appears that the best will be the lowest situation, or that at General Lambton's Rock. But for a double arch bridge M^{r}. Richardson has made it probable that the middle situat^{n}. will be best —

As to the choice between a stone bridge, & a wooden one,[4] I think the observations & remarks of M^{r}. Richardson are very judicious, & deserving of the mature consideration of the Commissioners. His opinion goes much in favor of the wooden one, as to the article of expence; for in no other point of view can there be any comparison between them. His reasoning I believe however is only meant to be applied in the case of a bridge of 2 arches, as I apprehend it would not answer for the single large arch. In favor however of a stone bridge it may be remarked that in the mind of every person there is a predilection in its favor, & a hope that the tolls or revenues may be made equal to the expence, & that hereby the necessity of rebuilding & often repairing a wooden one is obviated, as well as the continual fear & even dread of its falling, which the mind can never wholly divest itself of; & that this less stable & less secure tenure of it, make it necessary that the interest of the money advanced should be the higher, which has the effect of making the difference in point of expence less than in the proportion as stated by M^{r}. Richardson. Add to this, that long experience is against the use of wooden bridges; as, in the gradual improving state of things, we often find an old wooden bridge removed, & a stone one erected in its stead, but never the contrary; & persons living in the neighbourhood of a wooden bridge are always dissatisfyed with them, & looking forward to the time when it shall be changed for a stone one. There is also another

[a] Altered from ?*betwixt*.

circumstance which greatly affects the pecuniary comparison between the two sorts of bridges, & that is the changing value of money in a course of Years, which operates greatly against the wooden bridge, or any thing that will require frequent repairs or rebuilding; for the same work will require much more money to perform it 20 or 50 Years hence, than now; so that all the future repairs & rebuildings ought to be estimated much higher than they are at present, probably so high as double after the term of 50 Years to what they would be now —

There are some smaller inaccuracies in M[r]. R's Estimate of expences, which I have corrected & marked with a pencil, as you will see in his papers, which I have returned, as I do not know but You may want them, having divided them to bring them into the weight allowed to franks —

As to the more minute details of plans & elevations, & the mode of moving[a] the works, they will come better in hereafter, when the situation & n[o]. of arches are determined upon, & the whole business is in a more advanced state —

In the mean time, I would advise the greatest circumspection to be used in the choice of the architect, to acquire a good knowledge & assurance of his Character & abilities, & last of all perhaps to take some such security for the standing of the bridge as that mentioned above. I would advise to leave the contract open as long as possible, to receive the projects & estimates of all candidates, giving them all an equal & fair chance, & perhaps even advertising for projects or proposals, allowing each candidate a small & moderate sum agreed upon for travelling expences, if he come from a distance to view the situation; but giving every one to understand that he is to have no allowance for his drawings or projects, to prevent them from claiming & recovering such charges as they may chuse to make under the pretence that you desired them to make such drawings, &[c]. without specifying any express condition or agreement; or else allowing the same indulgences to all candidates alike — Wishing the best success to the business,

I remain, Sir,
Your most obedient
humble Servant
Cha: Hutton —

To
R. Burdon Esquire

[a]Source is unclear.

Notes

Rowland Burdon (1756–1838), gentleman, was a native of Newcastle. From 1790 to 1806 he represented Durham County in Parliament, and in connection with the improvement of roads he was involved with the planning and building of the cast-iron bridge at Sunderland. As well as Hutton he consulted John Soane, John Nash, Robert Mylne, John Smeaton, and Joshua Walker about the project: Skempton and Chrimes, *Civil Engineers*, 102–3. Hutton had been interested in the theory of bridge-building since the collapse of Newcastle bridge in 1771, in response to which he wrote his 1772 *Bridges*. See the postscript to Letter 67 for the sequel to Hutton's advice.

[1]I have not identified Mr Mackintosh, who I think is unlikely to be the engineering contractor Hugh McIntosh (1768–1840: according to Skempton and Chrimes, *Civil Engineers*, 422, he did not work on a bridge project until 1814), nor Mr Richardson.

[2]The plan with one arch was eventually adopted.

[3]The 'arch of equilibration' appeared in Hutton's own work on the design of bridges, denoting the intrados of a bridge in which the forces at each point were balanced. Thus a cycloid or a catenary could be an arch of equilibration, for the right extrados; conversely, for a given extrados (such as a horizontal line) the intrados which was an arch of equilibration could be found by an algebraic process. See Hutton, *Bridges*, 18–44 and Cecchi, 'The "Arch of Equilibration" of Charles Hutton (1772)'.

[4]The bridge as built was of cast iron; in fact only the second large iron bridge put into use in Britain. See Hutton, *Tracts* (1812), vol. 1, pp. 146–7.

Letter 60.
John Playfair to Charles Hutton, 25 August 1792

London, Wellcome Collection, MS 5270 no. 44: ALS.

Edin[r]. 25th Aug[t]. 1792.

Dear Sir

Since I wrote you last I have made some progress in the Life of James Gregory,[1] and hope to be able to send it you against Nov[r]. or thereabouts. What cheifly obstructs my progress at present is the want of a complete set of his Works, which is not to be found in our public Libraries. I must therefore take the Liberty of requesting, as I think you have nearly printed off the whole, that you will send me down a copy of as much as is actually executed.[2] This will save me a great deal of trouble in collecting his pieces from different quarters, and will enable me to adjust all the references I may have occasion to make to the new Edition of his Works. The Book may be sent directed to me[a] Nicholson Street Edin[r], by the Mail Coach.

[a] *deleted*: Nicholson.

I hope you will excuse this trouble, and you will lay[a] me under an additional obligation if you will at the same time let me hear that you are well, and that the Phil & Math. Dictionary[3] is in forwardness, both of which Circumstances very much interest your sincere Freind & humble serv[t]
John Playfair

Notes

For John Playfair see note to Letter 19.
[1]See Letter 39.
[2]The implication seems to be that Playfair believed Hutton was involved with the printing of the complete works of the mathematician James Gregory (1638–75). No such edition appeared around this time, although Maseres, *Scriptores Logarithmici*, vol. 2, pp. 1–19 contained extracts from certain of his works.
[3]That is, Hutton, *Dictionary*; see Letter 62.

Letter 61.
CHARLES HENRY VIGNOLES to CHARLES HUTTON, 3 June 1793

Portsmouth, History Centre, Vignoles Papers, 1072A/App. 4: ALS (our source).
Also printed in Vignoles, *Infant Ensign* as Appendix 4.

Youghall 3[d] June 1793

D[r] Sir

I take the earliest Opportunity of congratulating you on the Birth of a Grandson,[1] my Beloved Camilla I was obliged to part with leaving her with the Good Familly I informed you off in my last which I suppose you have long since Received & I have just Received a Letter from M[r] Blacker informing me she was safely delivered on Friday Evening twenty minutes before seven OClock, her time was tedious but thank God she had a good Nights Rest & is as well this morning as can be expected. I am in Great haste as I sett off tomorrow morning for Carlisle Fort near the Cove of Cork where I am to Remain untill Further Orders[.] I can only Add my Love & congratulations to all the Familly & am
D[r] Sir your most Affec[t] Son
C.H. Vignoles

---

[a] *altered from*: lie.

Note

For Charles Henry Vignoles see note to Letter 49.

[1]Charles Blacker Vignoles (1793–1875), civil engineer, was the child of Charles Henry Vignoles and Camilla Hutton (for whom see note to Letter 16); after their death Charles Hutton took him into his household. Hutton and Vignoles quarrelled in 1813 and broke irrevocably in 1817 after Vignoles contracted a clandestine marriage and departed for America. Vignoles went on to have a distinguished career as a civil engineer; he was eventually elected to Fellowship of the Royal Society and knighted: ODNB; Vignoles, *Life of Charles Blacker Vignoles;* Vignoles, *Infant Ensign*; Vignoles, *Romantic Engineer;* Wardhaugh, *Gunpowder and Geometry*; Letters 66, 78, 93, 100, and 111.

Letter 62.
CHARLES HUTTON to NEVIL MASKELYNE, 19 December 1793

Cambridge, University Library, RGO 4/187/18: ALS.

The whole text has been roughly crossed through, and the reverse of the sheet is covered with mathematical notes in Maskelyne's hand.

Woolwich Com^{n}. Decr 19, 1793.

Dear Sir,

I thank you for complying with Mr Robertson's[1] request, which I inform him of by this Day's post. Thank you also for your intention as to Mason's Tables,[2] of which I should like to have a Share in the transformation, &c.

I have just got to the end of the Alphabet with my Dictionary,[3] & have sent a few sheets to the press of that work. The bulk of the Copy, now it is got all together, has a formidable appearance, making more than 3 reams of folio sheets of paper, all written on one side, with my own hand.[4] And I am thankful to God it is so far out of my hands. — I hope we shall meet tomorrow at the Club[5] & on Saturday at 10 o'Clock in the Warren.[6] There are 19 young Expectants, but most of them only through Euclid & my book on Mensuration,[7] with 3 or 4 only just entered on Conic Sections, & a little Algebra, as far as Quadratic Equations.

Mrs H. (who is not very well) & my Daughters[8] join in respects with,
Dear Sir,
Your most obed^{t} Servt
Cha. Hutton.

Notes

For Nevil Maskelyne see note to Letter 13.

[1]For Abram Robertson see note to Letter 53.

[2]The astronomer and mathematician Charles Mason (1728–86), best known as one of the measurers of the Dixie Line, produced improved versions of the lunar tables of Tobias Mayer under the patronage of the Board of Longitude from 1772 onwards; they were published in Mason, *Mayer's Lunar Tables*. It appearing that the precession of the equinoxes (the secular movement of the earth's axis of rotation) was slightly slower than hitherto believed, it became necessary to introduce a correction; an essay by Maskelyne in the 1798 *Nautical Almanac* (printed in 1793, but the essay is dated December 1791) explained the situation and described how to correct for its effects on the data in the *Almanac*. It is not clear what Maskelyne's 'intention' was with respect to the tables at the date of this letter; they were not, in the event, republished during the 1790s.

[3]Hutton, *Dictionary*; Hutton would state in 1795 (Letter 70) that he had given the project 'close attention' for '10 or 12 years', in other words since about 1783–5.

[4]Compare Letter 67, where Hutton stated that the manuscript included 'some copying' by his youngest daughter Charlotte, aged fifteen at this time.

[5]For the 'Friday' Club see note to Letter 20.

[6]That is, at the Royal Military Academy at Woolwich. The implication is that Maskelyne was to assist at an examination of cadets; one indeed took place on 21 December, although the surviving records of the Academy do not mention Maskelyne: Jones, *Records of the RMA*, 37.

[7]That is, Hutton, *Mensuration*; at this time the first four books of Euclid's *Elements* seem to have been expected: Jones, *Records of the RMA*, 23, 38.

[8]Isabella and Charlotte (see notes to Letters 16 and 67) were certainly at home at this time; it is not certain when Eleanor (see note to Letter 16) married and left home.

Letter 63.
LEWIS EVANS to CHARLES HUTTON, 29 April 1794

Oxford, Museum of the History of Science, MS Evans 31: draft.

Little Bedwin April 29th. 1794

Dear Sir

I, at length, send you my packet of solutions which I hope will meet your approbation. I am your
most obliged servant
L.E.

...

Note

For Lewis Evans see note to Letter 17.

Letter 64.
LEWIS EVANS to CHARLES HUTTON, 24 May 1794

Oxford, Museum of the History of Science, MS Evans 31: draft.

Little Bedwin
May 24th. 1794

Dear Sir

Upon mature consideration of my new question,[1] it appears to me to be more abstruse than I at first apprehended, on account of the difficulty to obtain the dimensions of the oblique elliptical cone generated by the proposed position of its axis to the horizon. Should you think the question of sufficient importance to insert it, you may depend on my utmost exertions to give it a solution. It may not be amiss however to observe, that the angle of inclination I have given the axis of the glass is too small; instead of 55°[a] I would beg to say 65° in which case none of the wine will be spilt.

If you please, you may mention in a note my intentions of publishing the work I spoke to you of[b] last year, as soon as a sufficient number of subscribers can be ascertained for defraying the expence of publication.[2] I am Dear sir

your faithful & affectionate
humble servant
L. Evans

Notes

For Lewis Evans see note to Letter 17.

[1]Evans referred to a question included with Letter 63 and which eventually appeared thus in the 1796 *Ladies' Diary* (p. 48):

> A gentleman having drunk half a conical glass of wine, the diameter of which being $1\frac{1}{2}$ inch, and depth 2 inches, inclined its axis to an angle of 65° with the plane of the horizon; Query the area of the surface of the wine in that position?

---

[a] *altered from*: 65°.
[b] *altered from*: of to you.

Evans had once again (compare Letter 41) failed to provide a solution at the time of submission; but see Letter 65, dated just three days later, for the happy sequel.

[2]In the 1795 *Ladies' Diary* at p. 34 there appeared a note that Evans

> intends (as soon as a sufficient number of subscribers can be ascertained for defraying the expence) to publish, The Mechanical Construction of Cata-dioptric Telescopes of the Cassegrain, Gregorian, and Newtonian form, &c; with a description of the various tools and machines used in their fabrication.

It does not seem that any such work was ever printed.

Letter 65.
LEWIS EVANS to CHARLES HUTTON, 27 May 1794

Oxford, Museum of the History of Science, MS Evans 31: draft.

May 27th. 1794.

Dear Sir

Since I had the pleasure of addressing you on the subject of my question[1] I have the satisfaction to inform you that I have hit upon a method of solution, which is as follows.

...

you may rely on my sending[a] the numerical calculation in proper time. I am Dr. Sir,
most sincerely yours
L. Evans

Note

For Lewis Evans see note to Letter 17.

[1]This was the question included with Letter 63 and mentioned in Letter 64, where Evans had promised his 'utmost exertions' in search of a solution.

---

[a] *deleted*: you.

Letter 66. M. COURTOIS to CHARLES HUTTON, 15 December 1794

Portsmouth, History Centre, Vignoles Papers, 1072A/App. 10: ALS (our source).
Also printed in Vignoles, *Infant Ensign* as Appendix 10.
Also printed in 'The Vignoles Snuff-Box', 47.
This letter was brought to Hutton by his son Henry along with his grandson Charles Blacker Vignoles, reaching him in August 1795: see Letters 71 and 93. The news it contained had reached England, in a somewhat garbled form, much earlier: in October 1794 the death of Charles Henry Vignoles, Camilla Vignoles, and (incorrectly) Charles Blacker Vignoles was reported in the *Gentleman's Magazine* (p. 957).

au Port de La Liberté, ci devant Pointe à pitre Isle Guadeloupe le 25.
frimaire L'an 3. de La République francoise
15. décembre 1794. vieux stile.

M. de Vignolles Capitaine au 43e. Régiment[1] est mort à La Pointe à pître le 8. Juin 1794. Madame de Vignolles son épouse[2] ne lui a survécu que de deux jours elle mourut le 10.

Cette aimable femme qui m'avoit souvent entretenu de son tendre attachement pour son pere & son frere, sentant son dernier moment s'approcher, vouloit se donner la satisfaction de leur écrire: elle demanda de l'encre et du papier, mais les forces lui manquerent: elle ne p[e]ut que me laisser leur adresse et me pria de leur témoigner de sa part tout le regrêt qu'elle avoit de mourir loin d'eux. Elle finit en me recommandant son enfant nommé Charles[3] alors âgé de treize mois qui déjà très malade ne paroissoit pas devoir lui survivre longtems. Je lui promis d'en prendre soin jusqu'à ce que le retour de la paix me permit de le renvoyer à sa famille. M Hutton son oncle[4] est venu le réclamer. ne pouvant le confier en des mains plus sûres, Je le laisse partir, espérant que sa santé bien rétablie le mettra à même de supporter les fatigues du voyage et que sa présence adoucira le chagrin que M. le Docteur Hutton va ressentir à la nouvelle de la perte de sa fille chérie.

Courtois.

Doctor Hutton
Woolwich Common.
Kent
England

Captain Hutton
Royal English Artillery
Granada[a]

Notes

M. Courtois, benefactor of Charles Henry Vignoles and his wife and son, and co-owner of the firm of Courtois and Binbridge of Guadeloupe (Portsmouth, History Centre, Vignoles Papers, Letter 411/9), is not known apart from this and Letter 93, where some further details are given of the events of 1794.

[1]For Charles Henry Vignoles see note to Letter 49.

[2]For Camilla Vignoles, née Hutton, see note to Letter 16.

[3]For Charles Blacker Vignoles see note to Letter 61.

[4]For Henry Hutton see note to Letter 12. Having entered French-held Guadeloupe in search of his orphaned nephew, he found the British garrison at Berville in such a poor state that he felt compelled to offer his services to the commanding officer; he lost an eye in the ensuing action on 30 September and was taken prisoner. Subsequently he reached Courtois and Charles Blacker Vignoles presumably in December 1794 and returned to England with him and his nurse: their arrival is reported in Letter 71. See also the obituary of Henry Hutton in the *Gentleman's Magazine* (December 1827), pp. 561–2.

Letter 67. CHARLES HUTTON to ROBERT HARRISON, April–August 1795

Printed in Melmore, 'Some Letters', 79–80.
Date from internal evidence.

WOOLWICH COMN.

MY DEAR FRIEND,

We both (Mrs. H. & myself) thank you a thousand times most sincerely & affectionately for your kind remembrance of us, & especially of our dear lovely angel, for ever to be lamented.[1] O my friend never did I think it possible I could feel such severe affliction! And sure never parents had so much

[a]The two addresses appear at the top of the paper in a weak hand, evidently that of Camilla Vignoles.

cause. It is indeed impossible to express her excellence. This mournful event has quite altered all my future views & plans. I have begun a manufactory & otherwise engaged in plans of business.[2] Now I am hastening out of them with all speed, in order to return to a retired & recluse situation for the gloomy remainder of a lingering existence.

I am sorry, my friend, on several accounts that of late years you have written so little, as your writing if your health would permit, might be so cheering to your friends, & so useful to the public. But since you last left us, my dear friend, many has been the heavy ream of paper I have been obliged to write. The Dictionary alone (not half my occupation) has amounted to upwards of 3 reams of folio paper, all written, & mostly by my own hand, except some copying by the dear hand of my departed angel.[3] An immense task. I now however have reached the end of it, & I am repaid by the very favourable manner in which it is received by the public.[4] It will add much to my happiness if it yield any amusement to my friend in the perusal, for whose sake I am sorry it is printed in so small a type, much against my will. Two parts are now published, which you receive with this, making the 1st vol. but wanting the preface, which will accompany the fourth or last part. These I order to be sent by the proprietor Mr. Johnson.[5] Along with them I send a present of one of the Copies of the Logarithms presented to me by Mr. Baron Maseres,[6] along with his Compliments to you. There is another vol. of the Logs printing.

I remain, with sincere regard & affection,
My Dear friend,
Your obliged Servant
CHA. HUTTON.

P.S. I send also 2 Copies of the pamphlets relating to the disputes in the R. Soc.[7] one for yourself, & the other to bestow where you best may.

Mr. R. Burdon,[8] the member for Durham, has treated me very shabbily. The project for the Stone Bridge at Sunderland he referred to the Consideration of myself & Mr. Soane an eminent architect.[9] On this business I had a vast deal of trouble, many calculations, & a great many journeys to London, & besides one's judgment may be supposed worth something. In consequence of our report & advice the Commissioners saved a great many thousand pounds. When all was over, we expected they would make us a handsome present, as the genteelest way. Instead of that Mr. B. desired to know what quantity of trouble we had had, and how much we rated it at, &

would not be put off, but insisted upon our naming a certain sum. And, as if this were not degrading enough, after we had thus been brought to name 150 gs each for our Judgment & opinion, he refused to give us more than 50 for the whole!

Notes

For Robert Harrison see notes to Letter 9. The letter can be dated from its reference to the publication of the second 'part' of the *Dictionary* (but not yet the third), which according to the schedule of publication advertised in various London papers during January 1795 would place it between late April and perhaps late July of that year. It seems in any event very likely this letter was written before Letter 71 to Harrison, dated 13 August 1795.

[1]Charlotte Matilda Hutton (1778–94), the only child of Hutton and his second wife Margaret Ord (see note to Letter 31); she died on 24 September 1794: obituaries in the *Gentleman's Magazine* (October 1794), pp. 960–1; in the *Hibernian Magazine* (1794), pp. 477–8; and the account of her life in Johnson and Exley, 'Hutton, Charlotte'.

It is strange that Hutton did not mention the death of Charlotte's sister Camilla, which took place in June 1794 and had been reported in the London newspapers the following October: see Letters 66 and 71. Compare Hutton's extremely brief reference to Camilla's death in Letter 71.

[2]More than one early account of Hutton's life mentioned that he had been involved in the manufacture of bricks in connection with the houses he erected on Woolwich Common: Bruce, *Memoir*, 20; 'Public Characters', 122.

[3]Compare the account of the '3 reams' of the *Dictionary* in Letter 62. The obituary notice for Charlotte Hutton in the *Gentleman's Magazine* (October 1794), pp. 960–1 confirms (p. 960) that she acted as 'amanuensis and assistant on all occasions' for her father.

[4]Reviews of the *Dictionary* were in fact decidedly mixed, and it is not clear exactly what Hutton had in mind here.

[5]Joseph Johnson (1738–1809) was a prominent and successful bookseller in London from the 1760s: ODNB. Hutton's contract for the *Dictionary* was with Johnson alone, although G.G. and J. Robinson were also named in the imprint to the first (1795) edition: University of Cincinnati MS Q121.H93 1786, Memorandum of agreement between Charles Hutton and Jos. Johnson, 20 May 1786.

[6]That is, volumes 1 and 2 of Maseres, *Scriptores Logarithmici*, both of which had been published in 1791; volume 3 did not appear until 1796. For Francis Maseres see note to Letter 7. The assertion that Maseres had 'presented' copies of the *Scriptores Logarithmici* to Hutton seems to indicate that Playfair (see Letter 39) was wrong to consider Hutton a joint editor of that work.

[7]By 1815 Hutton still possessed copies of [Wolcot], *Peter's Prophecy*, and multiple copies of Maty, *Authentic Narrative* and *An Appeal*, making it likely that the material he sent to Harrison included at least the last two (see *A Catalogue*). The 'Dissensions' of 1784 were also the subject of: *Supplement*; *An History*; Kippis, *Observations*; [Strange], *Canons of Criticism*; and perhaps of 'Simon Snip', *Puppet Show*.

[8]For Rowland Burdon and his Sunderland bridge project see Letter 59 and notes there.

[9]Sir John Soane (1753–1837), with whom Burdon had in fact studied in Italy: ODNB; Skempton and Chrimes, *Civil Engineers*, p. 102.

Letter 68.
THOMAS COLLINSON to CHARLES HUTTON, 1795–6

Printed in Hutton, *Dictionary* (1795), vol. 2, p. 730 (our source).
Reprinted in part, with slight changes of orthography, in Hutton, *Dictionary* (1815), vol. 1, p. 194.
Date from internal evidence.

Turning over the leaves of your late valuable publication [...] part 1. of the Mathematical and Philosophical Dictionary, I observed under the article *Automaton*, the following: 'But all these seem to be inferior to M. Kempell's[1] chess-player, which may truly be considered as the greatest master-piece in mechanics that ever appeared in the world;' [...] So it certainly would have been, had its scientific movements depended merely on mechanism. Being slightly acquainted with M. Kempell when he exhibited his chess-playing figure in London, I called on him about five years since at his house at Vienna; another gentleman and myself being then on a tour on the continent. The baron (for I think he is such) shewed me some working models which he had lately made—among them, an improvement on Arkwright's cotton-mill, and also one which he thought an improvement on Boulton and Watt's last steam-engine. I asked him after a piece of speaking mechanism, which he had shewn me when in London. It spoke as before, and I gave the same word as I gave when I first saw it, *Exploitation*, which it distinctly pronounced with the French accent. But I particularly noticed, that not a word passed about the chess-player; and of course I did not ask to see it.—In the progress of the tour I came to Dresden, where becoming acquainted with Mr. Eden, our envoy there, by means of a letter given me by his brother lord Auckland, who was ambassador when I was at Madrid, he obligingly accompanied me in seeing several things worthy of attention. And he introduced my companion and myself to a gentleman of rank and talents, named Joseph Freidrick Freyhere, who seems completely to have discovered the *Vitality* and soul of the chess-playing figure. This gentleman courteously presented me with the treatise he had published, dated at Dresden, Sept. 30, 1789,[2] explaining its principles, accompanied with curious plates neatly coloured. This treatise is in the German language; and I hope soon to get a translation of it. A well-taught boy, very thin and small of his age (sufficiently so that he could be concealed in a drawer almost immediately under the chess-board), agitated the whole. Even after this abatement of its being strictly an automaton, much ingenuity remains to the contriver.—This discovery at Dresden accounts for the silence about it at Vienna; for I understand, by Mr. Eden, that Mr. Freyhere had sent a copy to baron Kempell: though he seems unwilling to acknowledge that Mr. F. has completely analysed the whole.

I know that long and uninteresting letters are formidable things to men who know the value of time and science: but as this happens to be upon the subject, forgive me for adding one very admirable piece of mechanism to those you have touched upon. When at Geneva, I called upon Droz, son of the original Droz of la Chaux de Fonds (where I also was).[3] He shewed me an oval gold snuff box, about (if I recollect right) 4 inches and a half long, by 3 inches broad, and about an inch and a half thick. It was double, having an horizontal partition; so that it may be considered as one box placed on another, with a lid of course to each box—One contained snuff—In the other, as soon as the lid was opened, there rose up a very small bird, of green enamelled gold, sitting on a gold stand. Immediately this minute curiosity wagged its tail, shook its wings, opened its bill of white enamelled gold, and poured forth, minute as it was (being only three quarters of an inch from the beak to the extremity of the tail) such a clear melodious song, as would have filled a room of 20 or 30 feet square with its harmony.—Droz agreed to meet me at Florence; and we visited the Abbé Fontana[4] together. He afterwards joined me at Rome, and exhibited his bird to the pope and the cardinals in the Vatican palace, to the admiration, I may say to the astonishment of all who saw and heard it.

Notes

Hutton stated in the accompanying text that the author of this and Letter 69 was the nephew of Peter Collinson, F.R.S. (1694–1768: ODNB). He was very probably the London banker Thomas Collinson (1727–1803), acquaintance of Joseph Banks and Benjamin Franklin among others, and correspondent on scientific matters: ODNB, *s.vv.* 'Robinson, Elihu (1734–1809), meteorologist and horticulturist' and 'Brown, Henton (1697/8–1775), Quaker minister and banker'; also Wigham, 'A Bag of Old Letters'.

The letter was evidently sent between the publication of the first part of the *Dictionary* in January 1795 and the appearance of volume 2 containing parts 3 and 4, in the appendix to which the letter was printed, in February 1796.

[1]The reference was to the notorious fraud the chess-playing automaton, sometimes called the 'Mechanical Turk', exhibited by Baron Wolfgang von Kempelen de Pázmanánd (1734–1804: *Neue Deutsche Biographie*, vol. 11, p. 484) around 1770.

[2]Collinson referred to Joseph Friedrich, Freiherr von Rackwitz (1744–1818), and von Rackwitz, *Ueber den Schachspieler des herrn von Kempelen*.

[3]The reference was to Pierre Jaquet-Droz (1721–90) and his son Henri-Louis (1752–91): *Dictionnaire historique de la Suisse*.

[4]This is likely to have been the natural philosopher and museum curator Felice Fontana (1730–1805), who worked in Florence from the 1760s.

Letter 69.
THOMAS COLLINSON to CHARLES HUTTON, 1795–6

Printed (extract) in Hutton, *Dictionary* (1795), vol. 2, p. 730 (our source).
Reprinted with slight changes of orthography in Hutton, *Dictionary* (1815), vol. 1, p. 194.
Date from internal evidence.

Permit me to speak of another Automaton of Droz's,[1] which several years since he exhibited in England; and which, from my personal acquaintance, I had a commodious opportunity of particularly examining. It was a figure of a man, I think the size of life. It held in its hand a metal style; a card of Dutch vellum being laid under it. A spring was touched, which released the internal clockwork from its stop, when the figure immediately began to draw. Mr. Droz happening once to be sent for in a great hurry to wait upon some considerable personage at the west end of the town, left me in possession of the keys, which opened the recesses of all his machinery. He opened the drawing-master himself; wound it up; explained its leading parts; and taught me how to make it obey my requirings, as it had obeyed his own. Mr. Droz then went away. After the first card was finished, the figure rested. I put a second; and so on, to five separate cards, all different subjects: but five or six was the extent of its delineating powers. The first card contained, I may truly say, elegant portraits and likenesses of the king and queen, facing each other: and it was curious to observe with what precision the figure lifted up his pencil, in the transition of it from one point of the draft to another, without making the least slur whatever: for instance, in passing from the forehead to the eye, nose, and chin; or from the waving curls of the hair to the ear, &c. I have the cards now by me, &c, &c.

Note

For Thomas Collinson see note to Letter 68, which this letter evidently followed.

[1]For Henri-Louis Droz see note to Letter 68; he had a workshop in London from the 1770s.

Letter 70.
Charles Hutton to David Stephenson, 7 February 1795

Cambridge, University Library, White b.8: a single page advertisement for Hutton, *Dictionary*, with ALS attached (our source).
Facsimile printed in Wallis, 'Newcastle mathematical libraries', 10–11.

Woolwich, Feb. 7, 1795.

Dear Sir,

On the present occasion give me leave to trouble you with one of the printed proposals for my Dictionary, & to desire the favour you will be so good as communicate the same to your scientific friends, & particularly to the members of your Literary & Philosophical Society.[1] — I hope the Society is in a good & flourishing way. By my close attention to the work of this Dictionary for the last 10 or 12 years, I have been prevented from a communication with my friends on literary matters, which I shall ⟨wish⟩ to resume as soon as all the printing of it is over. — I hope your father,[2] as well as your self is well. I desire you will present my respects to him, as well as to Mr Newton & Mr Anderson[3] when you see those gentlemen, & believe me

D^{r} S^{r}

Your obed^{t} hble Se[rv]^{t}

Cha. Hutton.

Mr David Stephenson
Architect
Newcastle upon Tyne

Notes

David Stephenson (1757–1819) of Newcastle was by the beginning of the nineteenth century the leading architect in the north-east of England: Welford, *Men of Mark*, vol. 3, pp. 434–7; ODNB *s.v.* 'Dobson, John (1787–1865), architect'.

[1]The Newcastle Literary and Philosophical Society was founded in 1793 by a committee of four, of whom Stephenson was one. It developed a library and a programme of lectures over the following years, and Hutton took a keen interest in its affairs. See also Letters 86, 104, 106, 115, 112, 117, 129, and 130. When the Society's new building was opened in 1822 Hutton was toasted (in his absence) at the celebratory dinner; he subsequently left Gahagan's bust of himself to the Society: Report in *The Newcastle Courant* (7 September 1822); The National Archives, PROB 11/1666/375, fol. 297^{v}: Will of Charles Hutton.

[2]David Stephenson's father was John Stephenson, house carpenter, of Newcastle: Welford, *Men of Mark*, vol. 3, p. 434. Since the son was in his teens when Hutton left the north-east, it was presumably with the father that he was primarily acquainted.

[3]'Mr Newton' was almost certainly William Newton (1730–1798), architect, of Newcastle; he was active during Hutton's time in the city, and in 1795–6 worked with David Stephenson on the remodelling of Newcastle exchange and guildhall. 'Mr Anderson' was possibly George Anderson (*c.* 1726–98), for whom see note to Letter 9.

Letter 71.
CHARLES HUTTON to ROBERT HARRISON, 13 August 1795

Printed in Melmore, 'Some Letters', 80–81.

WOOLWICH, *Aug.* 13, 1795.

MY DEAR FRIEND,

I really thought you had been possessed of my Logarithms & Conic Sections, & Tracts.[1] I send orders now to Mr. Johnson to supply all these (none of which are my property)[2] & to send them through Mr. Ritson[3] along with the 3d part of the Dictionary. The two vols of Scriptores Log[4] please to bestow any where that you please. Thank you for the list of Errata in the Dictionary, & pray be so good as to continue the collection of them.

I have just received from the West Indes, my son, with the loss of his right eye by a musket shot; & along with him, my grandson, the son of my daughter Camilla, who with her husband Capt. Vignoles died of the yellow fever in the West Indies; a most noble dashing boy of 2 years old.[5] There is one of the Errata, above mentioned, I cannot make out; it is this, [p 57 l A*r*ian];[6] be so good as explain it.

Mrs. Hutton is laid up in bed, with the Gout in both feet, but desires to be commended to you, along with

Your affectionate friend
and obliged servant
CHA. HUTTON.

Notes

For Robert Harrison see note to Letter 9.

[1]Hutton, *Tables*; Hutton, *Conic Sections*; and Hutton, *Tracts* (1786).

[2]For Joseph Johnson see note to Letter 67.

[3]Very probably the antiquary Joseph Ritson (1752–1803), best known for his collection of material relating to the legend of Robin Hood. A friend of Robert Harrison, born in Stockton and based during the 1790s in Durham, he also spent periods in London: ODNB.

[4]Maseres, *Scriptores Logarithmici*: volumes 1 and 2 (of six) had been published by this date.

[5]See Letter 66. It is not clear whether Hutton had any advance warning of the appearance of his son and the grandson whom he believed (see note to Letter 66) to be dead. This extremely brief and almost casual reference to the death of his daughter Camilla forms a striking contrast with the outpouring of grief for her (half-)sister Charlotte in Letter 67, and it may raise the possibility that Hutton and Camilla had parted on poor terms; see note to Letter 93.

[6]I cannot find anything in Hutton's *Dictionary* corresponding to this erratum.

Letter 72. JOHN LESLIE to CHARLES HUTTON, 14 October 1795

London, Senate House Library, MS 913B/3/1 (xv): ALS.
Sender reported in an endorsement in Hutton's hand.
Later endorsed 'Dr Olinthus Gregory's Sale, March 18, 1842'.

Largo, Fifeshire
14th Octr 1795.

Dear Sir

My experience of your open & friendly disposition encourages me to solicit your support on an occasion where a recommendation from you would very materially promote my interests. — The professorship of Natural Philosophy at St Andrew's is now vacant, & the right of electing a successor is vested in the members of the University. The professors have no immediate connexions that they can bring forward at present without the most flagrant indecency; nor do I hear of any considerable candidate who has yet appeared. The field being thus pretty clear, the professors will probably be induced to decide from general fame. They are in general conscious of their ignorance in matters of science, & will naturally look abroad to be directed in their decision. At the same time, several of the members regret the decline of the university, & desire to contribute somewhat to raise its reputation. Altho' I was educated at St Andrew's & am well known to the professors, they will be disposed to think much more favourably

⟨of me⟩, when they learn that I am held in some estimation by my English friends. I entreat therefore that you seize an early opportunity of consulting Dr Maskelyne on this subject.[1] Should you draw up a joint letter, I trust the doctor would add his signature. It might be addressed to the Revd Dr Joseph M^c^Cormack, principal of the united colleges of St Salvators & St Leonards, as head of the university. I must beg leave to add that any delay might prove fatal, as the session commences in a fortnight & it will be almost impossible to prevail with any person on such short notice to teach the class, without being appointed to the office. — My hopes of success are not very sanguine, but I should be wanting to myself, if let slip so fair an opportunity.

I have been much entertained with the parts of your dictionary that have reached me, & hope soon to peruse the whole with undivided attention.[2] — I desire my respectful compliments to Mrs Hutton, & am

Dear Sir

with the most sincere respect

Your most obliged humble ser^t^

[*signature cut out*]

PS I have likewise set other engines in motion — & calmly expect the event.

Dr Charles Hutton
Professor at the R. Military Acad
Woolwich

Notes

John Leslie (1766–1832) was a mathematician and natural philosopher: ODNB, DSB, MPHE. See Letter 73 for Hutton's response to this request for support, and also Letters 91 and 95 for further rounds of job-seeking on Leslie's part; he was eventually elected to a university chair at Edinburgh University in 1805, and to a chair of natural philosophy in 1819.

[1]For Nevil Maskelyne see note to Letter 13.

[2]In a letter to Ralph Griffiths (for whom see note to Letter 85) dated 4 December 1794, Leslie requested a copy of the Dictionary to review:

> I hope you will entrust me with Dr Hutton's Math. & Phil. Dictionary, which will appear about christmas . . . I believe his work is a ca[p]ital one.

A few weeks later on 23 December he pressed Griffiths for the book:

> I am particularly solicitous to carry the 1[st] vol. of Hutton's Math. Dictionary along with me. It is already printed & is to be published in parts; so that I have no doubt but Robinson would give it to you upon being desired.

(Oxford, Bodleian Library, MS Add. C. 89, fols. 196 and 197 respectively.) A note to the first passage, in another hand, reads 'Let Mr L. do it.', but the picky, lukewarm review of the *Dictionary* that eventually appeared in Griffiths' *Monthly Review* (February 1798, pp. 184–201) is, I think, unlikely to be Leslie's work.

Letter 73.
CHARLES HUTTON to JOSEPH MCCORMACK, 19 October 1795

London, Senate House Library, MS 913B/3/1 (xv): draft reply written on the cover of Letter 72.

Rev[d]. Sir,

I understand that the Professorship of Natural Philos. in the Univers[y] of St Andrew's is at present vacant. I humbly crave your forgiveness, Sir, for the liberty I take in troubling you with a letter in favour of a gent[n]. who, in my opinion, would very ably & reputably fill that important office, namely, Mr John Leslie, a gentleman who received his education in that Univers[y], & whose talents & extensive acquirements would do honour to any learned seminary in the world. Mr Leslie, Sir, is a gentl[n]. whom I have known for several years, & have often held conversations with him, with equal pleasure & profit, on[a] various branches of the math[l]. & philos[l]. sciences, & have perused[b] many of his ingenious compositions in that way, several of w[ch] have been printed in some very respectable publications.

I trust, Sir, I shall be credited, when I assure you, that this my representation is not intended merely to serve the interest of Mr Leslie, but that ⟨also⟩ of the sciences & education in general, to say nothing of the univers[y] in particular, as I have the strongest hopes that the benefit of his[c] ⟨great⟩ learning & abilities will be experienced in the highest degree wherever they may be employed. — I have the hon[r]. to be

[a] *altered from*: in.
[b] *altered from*: seen.
[c] *altered from*: his.

Rev. S^{r},
Your &c.
CH.

[*In another hand: Oct^{r}.*] 19, 1795.

To the Rev. D^{r}. Joseph McCormack, Principal of the University of St. Andrew's, Nth Britain.

Note

John Leslie wrote in Letter 72 that Joseph McCormack was 'principal of the united colleges of St Salvators & St Leonards', and thus academic head of the University of St Andrews; he is otherwise quite obscure. Hutton wrote in response to Leslie's request in that letter for a testimonial.

Letter 74.
ABRAHAM ROBERTSON to CHARLES HUTTON, 16 December 1796

Oxford, Museum of the History of Science, MS Museum 41, item 1: ALS.
Cover missing; recipient noted in an endorsement.

New College Lane, Oxford, Decr. the 16th. 1796.

Dear Sir,

I take the earliest opportunity of returning you my sincere thanks for your kind letter to the Duke of Portland[1] in my favour. On Monday last I had the honour and pleasure of being elected to the Professorship[2] without opposition. This success will fix my residence to Oxford, and I beg leave to assure you that it will give me a sincere pleasure to see you or any of your Family here.

I am, with great respect,
your most obedient and much obliged humble Servt.
A. Robertson.

Notes

For Abraham Robertson see note to Letter 53.

[1]William Henry Cavendish Cavendish-Bentinck, third duke of Portland (1738–1809), was at this time Chancellor of Oxford University; he was also prime minister in 1783 and again in 1807–9: ODNB.

[2]That is, the Savilian Professorship of Geometry, which Robertson held from 1 January 1797.

Letter 75. CHARLES HUTTON to NEVIL MASKELYNE, 27 February 1796

Cambridge, University Library, RGO 4/187/26:1: ALS.
The enclosures mentioned in the letter are not extant.

Woolwich Com^{n}. Feb. 27, 1796.

Dear Sir,

Inclosed are the two sheets concerning Simpson's Method for certain series,[1] ⟨which I have taken a copy of.⟩ He seems to have taken some pains with this kind of series, & has succeeded pretty well. — You had also some ⟨other⟩ sheets with specimens of very slow series, having all the terms positive, which I rather think you meant I was also to take with me, but if that was your intention we made a mistake, as I did not get them. — The prop. in Sterling,[2] which transforms the Series whose terms are all +, to a series having its terms alternately + & –, is the 7th. The same thing is also done in Cor. 4 pa. 65 Simpson's Dissertations.[3] And the transformation, by both these places, is this: If the given series be $a + bx + cx^2 + dx^3$ &c, it is transformed to $\frac{1}{1-x} \times (a - Dz + D'z^2 - D''z^3 + D'''z^4$ &c); where $z = \frac{x}{1-x}$, & $D = a - b, D' = a - 2b + c, D'' = a - 3b + 3c - d$, &c, being the Differential series, or the first of the 1st, 2d, 3d, &c, Differences of the other coefficients a, b, c, d, &c. — Now, if the first series $a + bx + cx^2$ &c, converge slowly <u>only</u> from the circumstance of the coefficients a, b, c, &c being nearly equal to each, but x be very[a] small, then the Differential or transformed Series $\frac{1}{1-x} \times a - Dz + D'z^2$ &c will converge pretty well, & the sum may be obtained

[a] *deleted*: large.

nearly by several methods, & perhaps best by the 2d paper in my Tracts, beginning at pa. 11.[4] But if x is also large, as almost equal to 1, then the powers of the quantity $\frac{x}{1-x}$ in the transformed series will diverge so fast as to render this series of no use for summation. To remedy this inconvenience it was that I gave another method for th[e] original slow series $a + bx + cx^2$ &c, in my 3d Tract, pa. 38;[5] b[y] means of which the series may be pretty readily summed, & by no other method that I know of. The example there take[n] at the bottom of pa. 41 is a pretty[a] trying one, & is found true to 7 figures; if only 3 or 4 figures had been desired, it would have been very easily done indeed by this method. — When you receive Mr Hellins's paper concerning this transformation,[6] be so good as send it to me to let me see it. I wish they will try ⟨it⟩ by this example $x + \frac{1}{2}x^2 + \frac{1}{3}x^3 + \frac{1}{4}x^4$ &c, when $x = \frac{9}{10}$, or $x = \frac{99}{100}$, or such like[b]. — N.B. There is an error in the signs in Simpson's book, viz, pa. 65 line 5, in the denominators, th[e]y should be $1 - x, \overline{1-x}|^2, \overline{1-x}|^3$ &c, instead of $1 + x, \overline{1+x}|^2, \overline{1+x}|^3$, &c.

I find I have the 1st vol. of your Observations[7] well bound & lettered, which you gave me. I have also two or three small parts of the 2d vol. stitched up separately, which I can return if you will send me the 2d vol. bound compleat, & also the 3d vol. if it is completed. I remain,

Dr Sr

Your most obed[t] Serv[t]

Cha. Hutton.

The Rev[d] Dr Maskelyne
Astronomer Royal

Notes

For Nevil Maskelyne see note to Letter 13.

[1]The allusion was to the prolific mathematician Thomas Simpson (1710–61). The 'two sheets' may have concerned the same passage cited later in the letter; if not, it is uncertain to which of Simpson's several publications about series Hutton referred.

[2]Stirling, *Methodus differentialis*, 40–3.

[3]That is, Simpson, *Mathematical dissertations*, 65.

[4]Hutton, *Tracts* (1786), 11–37.

[5]Hutton, *Tracts* (1786), 38–44.

[6]Most likely Hellins, 'A New Method', which as published took the form of a letter to Maskelyne.

[7]That is, Maskelyne, *Observations*, the four volumes of which bear the imprint dates 1776, 1787, 1799, and 1811.

---

[a] *altered from*: very.

[b] *altered from*: &c.

Letter 76.
NEVIL MASKELYNE to CHARLES HUTTON, 20 June 1796, 1797, or 1798

Cambridge, University Library, RGO 35/92: partial photocopy of ALS.
Year deduced from internal evidence.

Greenwich June 20. 179[a]

Dear Sir,

I submit to your judgmen[t] th[e] following sketch of what might be put in ?a note to the exceptionable assertion of [?Mr] Lalande with respect to the state of mathematics in England about 1764[b 1] & remain

Dear Sir,
Yours sincerely
N. Maskelyne

Had Mr. Lalande been as distinguished a mathematician as he is an Astronomer, he would not have thus under-rated[c] the merits of our principal British mathematicians of the time he alludes to, Lyons, Emerson, Landen, Waring;[2] the last of whom is still living, being the author of some of the greatest discoveries in Algebra; algebraic curved lines, infinite series, increments and fluxions.

Or take it as on the other side, or alter it as you please
Turn over

[*There is no photocopy of the second side of text; the verso of this photocopied sheet bears a note in biro: 'Maskelyne to Charles Hutton. / Orig. sold by John Wilson, Autumn 1984.'*]

Dr. Charles Hutton
Woolwich Common

---

[a] illegible.
[b] *altered from*: ?1768.
[c] *altered from*: it would have been difficult to account for his thus under-rating.

Notes

For Nevil Maskelyne see note to Letter 13.

The article of Lalande to which Maskelyne responded appeared in January 1796, while Edward Waring died in August 1798; the letter was therefore written in 1796, 1797, or 1798.

[1]For Jérôme Lalande see note to Letter 10. Maskelyne referred to the remark in Lalande, 'Notice sur la vie de Condorcet', 143, that in the year 1764 there was no first-rate analyst in England. Edward Waring (see note 2) responded to that remark in a letter to the editor of the *Monthly Magazine*, printed in May 1799 after Waring's death (Waring, 'Original letter of the late Dr. Waring'); Maskelyne's text was used as part of a note introducing the letter, thus:

> Had Lalande been as distinguished a mathematian [*sic*] as he is an astronomer, he would not have thus under-rated the merits of such eminent British mathematicians as Lyons, Emerson, Landen, and Waring, who all flourished at that period. Dr. Waring, who published his "*Miscellanea Analytica*" in 1762, and his "*Meditationes Algebraicæ*" in 1770, was the author of some of the greatest discoveries in algebra, algebraic curve lines, infinite series, increments, and fluxions.

It is not clear whether or not Maskelyne had this particular use of the text in mind when he wrote this letter.

[2]For William Emerson and John Landen see notes to Letters 9 and 14 respectively. Israel Lyons (1739–75) was a botanist and mathematical prodigy; like Hutton he had done calculating work for the *Nautical Almanac*; he was also a recipient of the patronage of Joseph Banks, and travelled on Constantine John Phipps' expedition towards the north pole (see note to Letter 22): ODNB; Hutton, *Dictionary* (1795), vol. 2, pp. 58–9. Edward Waring (*c.* 1735–98) was Lucasian Professor of Mathematics at Cambridge from 1760; it was to him that the Royal Society awarded the Copley Medal in 1784 in an effort to demonstrate its support for mathematical research: ODNB.

Letter 77.
CHARLES HUTTON to MARGARET BRYAN, 6 January 1797

Printed in Bryan, *Astronomy* (1797), xi (our source).
Reprinted with slight changes of orthography in Bryan, *Astronomy* (1799), xiii and Bryan, *Astronomy* (1805), xiii.

Woolwich, Jan. 6, 1797.

Madam,

I herewith return the ingenious MS. of Astronomical Lectures you favored me with the sight of, which I have read over with great pleasure; and the more so, to find that even the learned and more difficult Sciences are thus beginning to be successfully cultivated by the extraordinary and elegant talents of the female writers of the present day.

Should you, Madam, give to your friends and to the public to benefit by the publication of[a] these your learned and useful labours, I beg to have the honor of being considered one of the encouragers of so useful a work;

And am, with great respect,
Madam,
Your most obedient, and most humble servant,
Charles Hutton.

Note

The educator Margaret Bryan (*fl.* 1795–1816) ran a number of schools, in Blackheath, London and Margate during the period from the 1790s to the 1810s. In 1797 she prepared her lectures on astronomy for publication and submitted them to Hutton (with whom she had no prior acquaintance) for perusal. He provided both this letter of approval—'his avowed sanction and patronage', without which Bryan stated she 'could not have ventured to offer a work of Science to the public'—and also what she described as 'active services, benevolently exerted to procure me extensive benefit and suffrage'. This referred perhaps to publicity or to contacts within the publishing world; one of the booksellers named in the imprint was George Kearsley, for whom see note to Letter 83. Hutton's name also appeared among the long list of subscribers to the work. Bryan evidently regarded the incident as a turning point in her career. The volume achieved some success, and she went on to publish further textbooks on natural philosophy (1806) and astronomy and geography (1815). ODNB; Bryan, *Astronomy* (1797), x, xvii.

Letter 78.
CHARLES HUTTON to WILLIAM WINDHAM, 20 February 1797

Portsmouth, History Centre, Vignoles Papers, 1072A/App. 13: draft (our source). Printed in Vignoles, *Infant Ensign* as Appendix 13.

[a]The 1805 text has ... *give to your friends and to the public, to benefit by the publication of, these your learned and useful labours ....*

Woolwich Kent
20 Feb[y] 97

Sir

In laying before you[a] the enclosed Accompt of Losses and expences sustained by the family of my late Son in law and Daughter[1] I am aware that the ⟨Kings⟩ Regulation has not provided for the reimbursement of the Value of effects lost by Officers who are deceased and whatever distinction may have been judged proper in this Respect I hope the present case will[b] admit of some deviation from the Rule that has been adopted & I therefore beg[c] you will have the goodness to lay the case before the King with my humble request that His Majesty will be graciously pleased to Grant a Warrant for the Sum of £230 in[d] consideration of the Losses sustained by the[e] family of the late Cap[n] Vignoles 43[d] Reg[f], which I shall see is faithfully applied toward the provision and education of his Infant Son[g] now under my care and only 3 Years and 8 Months old —[2]

I have the Honor to be
Sir Your m. obed Serv[t]
Cha[s]. Hutton.

To The R[t] Hon[ble]
W. Windham
His Majesty's Sec[y] at War

Notes

William Windham (1750–1810), politician, was at this time Secretary at War in Pitt's administration: ODNB. It is not clear whether this letter was ever sent, nor, if so, whether its petition was successful. This halting beginning later became a warmer personal correspondence between Windham and Hutton: see Letters 87, 88, and 105.

[1]For Charles Henry Vignoles and Camilla Vignoles see Letters 49 and 66, and note to Letter 16.

[2]Charles Blacker Vignoles: see note to Letter 61.

---

[a] *altered from*: enclosing.
[b] *deleted*: be.
[c] Altered from: *beg leave to mention that* then *request.*
[d] *deleted*: part.
[e] *altered from*: my.
[f] *altered from*: my late Son & Daughter.
[g] *altered from*: their only child.

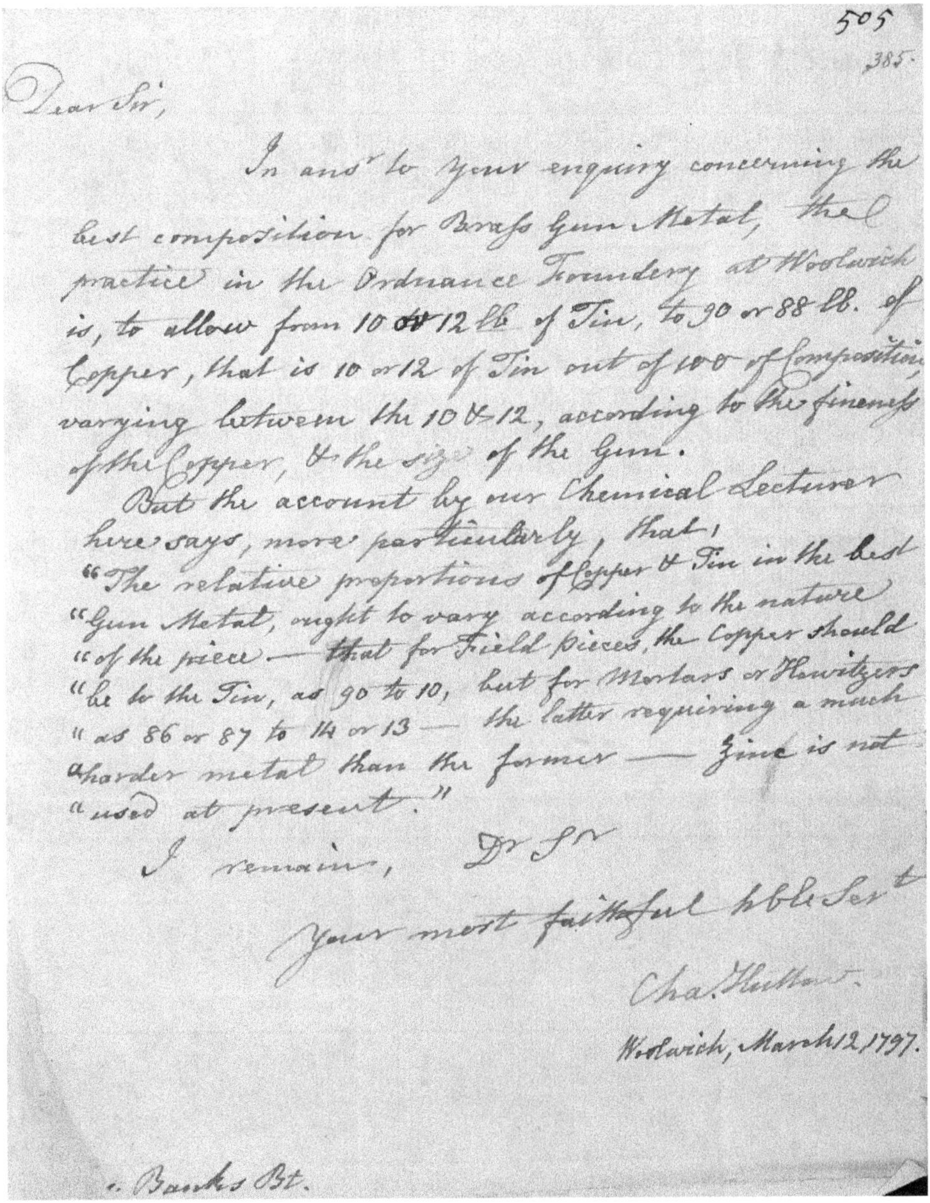

505
385.

Dear Sir,

In ans^r to your enquiry concerning the best composition for Brass Gun Metal, the practice in the Ordnance Foundery at Woolwich is, to allow from 10 or 12 lb. of Tin, to 90 or 88 lb. of Copper, that is 10 or 12 of Tin out of 100 of Composition, varying between the 10 & 12, according to the fineness of the Copper, & the size of the Gun.

But the account by our Chemical Lecturer here says, more particularly, that,
"The relative proportions of Copper & Tin in the best
"Gun Metal, ought to vary according to the nature
"of the piece — that for Field Pieces, the Copper should
"be to the Tin, as 90 to 10, but for Mortars or Howitzers
"as 86 or 87 to 14 or 13 — the latter requiring a much
"harder metal than the former — Zinc is not
"used at present."

I remain, Dr Sr

Your most faithful hble Servt

Cha. Hutton.

Woolwich, March 12, 1797.

.. Banks Bt.

Figure 3: Letter 79: Charles Hutton to Joseph Banks, 12 March 1797. British Library, Add. MS 8098, fol. 385r. ©The British Library Board.

Letter 79.
CHARLES HUTTON to JOSEPH BANKS, 12 March 1797

London, British Library, Add. MS 8098, fol. 385: ALS (our source).
Also printed in Chambers, *Correspondence of Banks*, vol. 4, p. 461.

Woolwich, March 12, 1797.

Dear Sir,

In ans[r] to your enquiry concerning the best composition for Brass Gun Metal, the practice in the Ordnance Foundery at Woolwich is, to allow from 10 or[a] 12 lb of Tin, to 90 or 88 lb. of Copper, that is 10 or 12 of Tin out of 100 of Composition, varying between the 10 & 12, according to the fineness of the Copper, & the size of the Gun.

But the account by our Chemical Lecturer here[1] says, more particularly, that,

"The relative proportions of Copper & Tin in the best Gun Metal, ought to vary according to the nature of the piece — that for Field Pieces, the Copper should be to the Tin, as 90 to 10, but for Mortars or Howitzers as 86 or 87 to 14 or 13 — the latter requiring a much harder metal than the former — zinc is not used at present."

I remain, D[r] S[r]
Your most faithful hble Ser[t]
Cha. Hutton.

[b]Banks Bt.

Note

For Joseph Banks see note to Letter 22.

[1]William Cruickshank (d. 1810/11) replaced Adair Crawford (1748–95) in 1796 as Chemical Lecturer at the Royal Military Academy; neither man seems to have published an account of the composition of gun metal, and Hutton presumably referred to a manuscript discussion by one or other man. Jones, *Records of the RMA*, iii.

---

[a] *altered from*: to.
[b]Paper damaged.

Letter 80. CHARLES HUTTON to E. COLLINS, 6 July 1797

London, British Library, Add. MS 33542, fol. 483: ALS.
Recipient identified in library catalogue and endorsement.

Woolwich
July 6, 97.

Sir,

I think I can now resolve the doubt about my form of Dr Halley's table of Mortality &c,[1] and it is simply as follows. You perceive that in my sett of tables, pa. 27 vol. 2 Dict, all the numbers are made to begin at 1000, & the ages begin at 0. But ⟨in⟩ Dr Halley's table, tho' the n^{os} begin at 1000, the ages begin 1 year; that is, his 1000 is the n^{o} of persons living at 1 year old, but mine are those living at 0, or then born. In my table therefore I reduced, by proper proportions, all his n^{os}. to the same beginning at 0, & to the same n° 1000 at 0. I am,

Sir
Your most obed^{t} Servt
Cha. Hutton.

Mr E. Col[lin]s[a]
at Mr Bentham's
Queen Square Place
Westminster.

Note

I cannot identify E. Collins. The Mr Bentham to whose house this letter was addressed was the philosopher Jeremy Bentham (1748–1832).

[1]The reference was to the table at Hutton, *Dictionary*, vol. 2, pp. 27–8, in the article on 'Life Annuities', which incorporated in modified form the data from the table in Halley, 'An Estimate of the Degrees of the Mortality of Mankind', 600.

[a]Paper is damaged.

Letter 81. CHARLES HUTTON to HENRY CAVENDISH, 17 November 1798

Chatsworth, Devonshire Collections: Cavendish MSS New Correspondence: ALS (our source).
Also printed in Jungnickel and McCormmach, *Cavendish*, 710–11.

Woolwich, Nov. 17, 1798.

Sir,

I have to return you thanks for a copy of your ingenious paper, on the Density of the Earth,[1] this day received from the Royal Society, which I suppose was ordered by yourself to be sent me. — Dr Maskelyne had before favoured me with a sight of the paper;[2] in which I perceive you bring out the density in question to be 5.48 times that of water. And, in the paragraph at the bottom of pa. 54, you remark on the difference between this conclusion, and & [*sic*] that of the Society's Experiment at Mount Schehallien, in the calculation of which I bore so great a share.[3] I wish that paragraph had been a little differently worded; because, as it stands, some persons may think you meant those words as some disparagement of that experiment, or a hint concerning some inaccuracy in the calculations or observations. If there be any such inaccuracies, it will be easy to detect them, as the particulars are detailed in so particular a manner, in the printed papers giving a Acco^{t}. of the observations & calculations, that any person may go over the whole numbers again.[4]

But it will be proper to remark, that the Schehallien experiment, as calculated in my paper, has not determined what is to be considered as the density of the earth, but only what is the ratio of that density to the density of the hill Schehallien, which it stated to be the ratio of 9 to 5. For the paper expressly states that we are ignorant as to the density of the hill, & consequently the real density of the earth also. All that was professed to be known, concerning the hill, was, that it is a mass of stone; but nothing as to the nature or density of that stone. I wish the Society had completed their experiment, by procuring a determination of the real density of the stone. And I wish they would now do it, that an accurate conclusion, as to the density of the earth, may be ⟨thence⟩ obtained.[5] For want of the real density of the hill, we barely made a <u>supposition</u>, that it is of the density of common stone, or $2\frac{1}{2}$, and on that supposition it followed, that the density of the earth would be $4\frac{1}{2}$. But there is great reason to expect that the density of the stone of the hill Schehallien, is much more than $2\frac{1}{2}$, probably more than 3, since the heavier stones, as granite, run as high as $3\frac{1}{2}$; and I expect that ⟨the⟩ density of the earth, when it comes to be determined from the real density of the hill, will turn out to be between 5 and 6, but probably nearer the latter number.[6]

I am, Sir,

Your most obed[t] hble Servant

Cha. Hutton.

7

Henry Cavendish Esq.
at the Royal Society
Somerset Place
[*in another hand*]: Through the hands of D[r]. Maskelyne

Notes

For Henry Cavendish see note to Letter 10.

[1]Cavendish, 'Experiments to determine the density of the Earth'.

[2]For Nevil Maskelyne see note to Letter 13.

[3]That is, the work reported in Maskelyne, 'An Account' and Hutton, 'An Account of the Calculations'. Cavendish had been a member of the committee that planned that experiment, and Hutton credited him with certain suggestions concerning the strategy of calculation.

[4]There was perhaps something disingenuous about the statement that 'any person may go over the whole numbers again': Hutton had called the work 'an immense labour' in Letter 12, and Bruce (for instance) reckoned the work more laborious than anything done by a single person 'since the preparation of logarithmic tables': Bruce, *Memoir*, 19.

[5]As Hutton explained in Letter 120, John Playfair would undertake a lithological survey of the Schiehallion site in 1801 for this purpose.

[6]Hutton was possibly guided by the fact that Isaac Newton had conjectured that the density of the earth lay between five and six times that of water: Hutton, 'The Force of Fired Gun-Powder', 783.

[7]Another hand copied at the bottom of Hutton's letter the final paragraph of Cavendish, 'Experiments to determine the density of the Earth':

> According to the experiments made by D[r] Maskelyne on the attraction of the hill Shehallien the density of the earth is $4\frac{1}{2}$ times that of water which differs rather more from the preceding [*Deleted*: exper] determination than I should have expected but I forbear entering into any consideration of which determination is most to be depended on till I have examined more carefully how much the preceding determination is affected by irregularities whose quantity I cannot measure

In Jungnickel and McCormmach, *Cavendish*, 712, this is printed as a draft of Cavendish's letter in reply to Hutton, but it is hard to see what sort of reply would consist (entirely) of a verbatim copy of part of a text Hutton had already seen, and I incline to think this was copied here for some other reason.

Letter 82.
FRANCIS MASERES to CHARLES HUTTON, 14 April 1801

London, British Library, Add. MS 28104, fol. 68: ALS.
The enclosure mentioned is not extant.

Inner Temple, april 14, 1801.

Dear Sir,

I have written to D^{r}. Hales[1] to inform him that you think the two passages mentioned by D^{r}. Magee[2] as erroneous, to be really so, and to deserve to be corrected. And I doubt not, he will soon send me over the proper corrections of them. I am glad to find that you agree with me in preferring Raphson's ⟨method of⟩ approximation to those of Newton, Halley, and De Lagny, though you think it somewhat inferiour to the Method of Trial and Errour, or of Double False Position, or what I call the Differential Method.[3] This last Method will always give[a] us a new value of the quantity sought that will be nearer ⟨to the truth⟩ than one of the two former values from which it is derived, as I have demonstrated in a note at the end of Tiphys Batavus, in the 4th volume of the Scriptores Logarithmici, page 271; but I believe it will not always be nearer to the truth than either of the ⟨said⟩ two former values; at least I have not been able to demonstrate that it will be so. Yet I believe that it will in most cases (though not perhaps in all,) be nearer to the truth than either of th[em.][b] I inclose herewith a Letter I have receivd for you [...] Dr. Andrew Mackay, of Aberdeen.[4] In most of [...] books of Astronomy I have looked into, the argument [...] deriving the conclusion "that the force of Gravity upon bodies decreases when the distances of the bodies from the Center of the Earth increase[c], and in the same proportion as the squares of those distances increase," from the velocity of the Moon in her orbit and the Periodick time of her Revolution, is grounded on Kepler's Observations that the areas described in

[a] *altered from*: gives.
[b] The paper is damaged.
[c] *altered from*: increases.

equal times are equal, and that the squares of the Periodick times are as the Cubes of the distances.[5] Now these things are true, and the reasonings grounded on them are just; but they seem to me not to be necessary in the present case; and therefore I have endeavourd to draw the same conclusion without their help.

I remain your most obedient and humble servant,

Francis Maseres.

To D^{r}. Hutton, at the Royal Military Academy at Woolwich in Kent.
By the penny-post.

Notes

For Francis Maseres see note to Letter 7.

[1]Presumably the Church of Ireland clergyman and scientific writer William Hales (1747–1831), a friend of Maseres: ODNB. The present passage probably referred to Hales, *Analysis fluxionum*, which was reprinted in 1804 in Maseres, *Scriptores Logarithmici*, vol. 5, pp. 91–204; see p. 91 there.

[2]Most likely William Magee (1766–1831), archbishop of Dublin in the Church of Ireland and better known as a theologian; he was in 1800 Professor of Mathematics at Trinity College: ODNB.

[3]Compare Hutton's remark on different methods of approximation in Letter 7, with the note there.

[4]Andrew Mackay (1758–1809), teacher of navigation, was at this time based in Aberdeen; although he refused an invitation to work at the Royal Greenwich Observatory, he subsequently moved to London and became acquainted with Hutton: ODNB, MPHE; see also Letter 98.

[5]See for example Newton, *Principia mathematica*, Book 1, Proposition 11, Problem 6.

Letter 83.
CHARLES HUTTON to GEORGE KEARSLEY, 13 July 1801

Printed in Despiau, *Select Amusements*, A2^{r}.

Woolwich, July 13, 1801.

SIR,

IN answer to your request, to examine and give you an opinion of a work, entitled "Philosophical and Mathematical Amusements, &c," being a translation from the French of Despiau, with additions, and improvements, I have no hesitation in declaring my opinion, that it is a very curious and ingenious work, comprising a great deal of useful matter in a small compass, and well adapted for communicating the knowledge of a great variety of interesting particulars, in a manner at once familiar, clear, and amusing. To young persons especially, and in schools, this work must prove peculiarly useful;[1] and the extensive tables of chances on all sorts of games, at the end of the work, must be singularly interesting to all such persons as attend or play at any kind of games, by exhibiting at one glance the chances or odds in a great variety of cases.

I am, Sir,

Your most humble Servant,

CHA. HUTTON.

To Mr. KEARSLEY, Bookseller, Fleet-Street.

Note

George Kearsley was the son of a better-known London bookseller of the same name (*c.* 1739–90); he was active in that trade from 1790 to 1813. Hutton's endorsement was mentioned on the title page of Despiau, *Select Amusements* and has occasionally given rise to the impression that he was the translator of that book: for instance obituaries of Hutton in the *European Magazine* (June 1823), pp. 483–7, at p. 487, and in the *Literary Gazette* (1 February 1823), pp. 75–6, at p. 75. There is no evidence for this.

[1]Compare the statement in a review of Hutton, *Recreations* (1814) in the *British critic* (1815), p. 418, that that work was of particular use to schools.

Letter 84.
CHARLES HUTTON to ?JOHN RENNIE, 10 February 1802

Kew, National Archive, PRO 30/9/131, fols 20–22: ALS.

Woolwich, Feb. 10, 1802

Sir,

In compliance with your request, in consequence of the honor done me by Mr Abbot,[1] to consider your designs for Iron bridges to be built over the river at Conway Castle, and over the strait near Bangor, I beg leave respectfully to remark as follows:

From the general view and consideration of your designs, there seems to be no doubt of the practicability of such erections, nor of the safety & permanency of them.

In regard to a comparison between the two proposed situations for the bridge over the Menai, there appear to be advantages & disadvantages on both sides. The situation at Inis y Moch seems to afford the more convenient passage to and from the bridge, & to admit of a design rather more graceful or sightly. But, on the other hand, the greater extent of the arch at that place will render it much less safe, and the great depth of water there will make it very difficult and expensive to erect a centering, on which to build the arch. So that, on the whole, the situation at the Swilly rocks may probably be the more safe and preferable.

In regard to your design for the Swilly bridge, it appears to me, that 100 feet, the thickness of the piers in the drawing, is far too much, both causing a needless expence, and rendering the design clumsy & ungraceful. — From the given weight and dimensions of the arch I have calculated the effect of its thrust or lateral pressure, and find that a peir of $28\frac{1}{2}$ feet thick, say 30 feet in round numbers, of solid square stone work, will be sufficient thickness to support the arch proposed, without however allowing any thing for the weight of the materials forming the road way over it, for computing which I have no data. And, as a hollow or tubular body is much stronger than a solid one, of the same quantity of materials, if the same 30 feet thick of solid matter be extended out so as to form a hollow pier of 60 feet thick, that is having 30 feet of solid masonry encasing 30 feet vacuity, it appears to me it will be quite sufficient for the arch & road way, and will look much better than the 100 feet pier.

The flanks of the arches seem to be too much loaded. If the arch be cut away in that part, so as to bring it to resemble an oval, it will look better, and bring it nearer to a true balanced arch.

The mass of dead wall adjoining the west end of the great arch at the Conway river, looks heavy. I would advise you to consider if it cannot be rendered lighter & be made more sightly, either by an arch there, or some other contrivance.

I am, with respect, Sir
your most obed^{t} Servt
Cha. Hutton.

W. Rennie
Stamford Street

Notes

I believe the 'W. Rennie' of the address was Hutton's mistake and that the recipient was the engineer John Rennie (1761–1821), perhaps best known for his work on canals. In 1821 Hutton signed his (successful) proposal for Fellowship of the Royal Society: ODNB, MPHE.

The letter discussed a proposal to bridge the Menai Strait, but in the event Rennie's high estimates of the cost meant that no bridge was built at this date; a suspension bridge by Telford eventually opened in 1826: see Skempton and Chrimes, *Civil Engineers*, 557. Evidently his 'shabby' treatment by Rowland Burdon (see Letter 67) had not discouraged Hutton from offering advice on such projects; see also Letter 87.

[1]I have not identified Mr Abbot.

Letter 85.
CHARLES HUTTON to RALPH GRIFFITHS, 1 May 1802

Oxford, Bodleian Library, MS Add. C. 89, fol. 167: ALS. Cover lost.

Woolwich, May 1, 1802

Sir,

Two days ago only I received your favour, in ans^{r}. to my letter to you of the 12th of April. It is stated that the writer of the offensive article[1] disclaims all motives of prejudice, ill-will, & intentional misrepresentation. But facts, Sir, speak better than words. No indifferent reader of that article can help ascribing the writer's account to some improper motives. Every person that I have consulted about it, is decidedly of the same opinion; and that the Reviewer meant, by insinuation & implication, to disparage the book and the author; that the account is fallacious & slighting; charged with sarcasm and misrepresentation. After such an article, & former ones of a like kind, it is in vain for the Reviewer to disclaim in words what is so strongly testified in his actions: like him "that casteth firebrands, arrows, & death, crying am not I in sport."

You express yourself satisfied with the Reviewer's declaration. It is very natural for you Sir to cover the matter the best you can; & you may be excusable. But still some consideration is due from you in such a case for the feelings of an author so outraged, without a knowledge of any offence. And, say what you can, the offensiveness of the reflections can never be explained away. — The observations respecting Speculative men[2] do still appear to be personal: they must have been introduced with that particular allusion & application, otherwise an exception would have been made in favour of the author of the book. Neither do those observations appear to be given as the assertions of others, like the quotation respecting "Men of the Pen", but as the reviewer's own thoughts, set in comparison & analogy with these.

— The opinion that my principles do not apply to the Iron Bridge,[3] was certainly unnecessary, officious, uncandid. — The remark that certain of my propositions are the same with those of Emerson,[4] is understood by all persons, that I have consulted, as ascribing plagiarism to their author, & not admitting of any other meaning. How can it be considered as "conveying a description of the propositions"? Is that a way to get a description of the prop^s in my book? Surely the book itself is the fittest place, & not another book published 4 years after it.

No, Sir, the allusions in the article are too obvious to be explained away; & you may be assured that the public understand them as I do. So that it is but too certain that an improper & prejudicial account has been given to them of my books. And the fact is not the less true that I am ignorant of the cause; whether it be thro' any particular pique & prejudice, or thro' a general propensity in Cambridge men to affect to despise, & endeavour to disparage, whatever comes not from their own society.

On the whole, Sir, I cannot help feeling that I have been treated injuriously, & that some notice ought to have been taken of it, to remove in part the erroneous idea the public have received from the Review. And tho' this has been refused me in the Review, I still think something ought to be done. So that, if you do not mean to make some use of my paper of observations, I will thank you to return it to me again by the post.[5] I would also ask as a favour of Dr Griffiths, that in case of any future publications of mine, he will not suffer them to be noticed in the Monthly Review at all,[6] which will much oblige his respectful humble Servant,

Cha. Hutton.

Notes

Ralph Griffiths (1720–1803) was the editor of the *Monthly Review*, in which a number of reviews critical of Hutton's works appeared during the years around 1800. The reviews were notionally anonymous but it seems to have been an open secret that the *Monthly*'s mathematical reviewer was Robert Woodhouse (1773–1827), Fellow of Caius College, Cambridge and an advocate of the importation of French mathematical methods into Britain: ODNB and see Letter 87. Evidently this letter was preceded by another from Hutton to Griffiths, enclosing a short article in response to the offensive review(s), and by Griffiths' response in which he declined to print that article.

[1]The 'offensive article' was a review of Hutton, *Bridges* (1801) in the *Monthly Review* 37 (March 1802), pp. 323–4.

[2]Woodhouse had stated of 'speculative men' that (p. 324):

> However inefficient in the actual formation of machinery, they have still a great inclination to instruct the mechanic and architect: but the instructions are generally slighted, and the suggestions treated as mere airy speculations, as the serious triflings of a theorist, claiming rank only among those unsubstantial systems which the pride of calculation is continually erecting, and which time and experience are constantly overthrowing.

[3]Woodhouse, p. 324 again:

> If we at all understand the model which has been exhibited to the public, the *cast iron bridge* will not derive its strength from the same principle which prevails in common arches.

[4]Woodhouse, p. 324 once more: 'Proposition 1st, Section 2d, of this tract [i.e. Hutton, *Bridges* (1801)] is the same with that of Emerson, p. 149. Miscellanies. — Prop. 3d and 4th are likewise nearly the same as those of Emerson.' For William Emerson see note to Letter 9; the work alluded to was Emerson, *Miscellanies*, namely article V on 'The construction of arches of bridges', pp. 148–79. As Hutton pointed out, this had appeared in print four years after the first (1772) edition of his *Bridges.*

[5]An article was subsequently published in the *Monthly Magazine* as Hutton, 'To the Editor of the Monthly Magazine'; it seems likely that this related, perhaps closely, to the 'paper of observations' mentioned here. In the event, Hutton was also allowed a briefer reply in the *Monthly Review* itself, Hutton, 'To the Editor of the Monthly Review'.

[6]In fact reviews of Hutton, *Recreations* and of the early volumes of Hutton, Shaw, and Pearson, *The Philosophical Transactions ... abridged* appeared in the *Monthly Review*, respectively in vol. 44 (May 1804), pp. 35–40 and vol. 47 (July 1805), pp. 284–6; they were lukewarm.

Letter 86.
CHARLES HUTTON to EDWARD BRUCE, 30 July 1802

Extract printed in Bruce, *Memoir*, 25.

Woolwich, July 30th, 1802.

I thank you for your offers of service, and shall take this opportunity to ask you the favour to procure and send me a copy of all the proceedings that have been printed by, or relative to, your Philosophical Society,[1] which I should like to have to bind them up regularly.

To Mr. Edward Bruce, Byker West House,
Newcastle on Tyne.

Note

Edward Bruce (*c.* 1774–*c.* 1806) was a Newcastle schoolmaster, the older brother of John Bruce whose acquaintance and correspondence with Hutton is better documented: see note to Letter 104 and Welford, *Men of Mark*, vol. 1, pp. 409–10.

[1]For the Newcastle Literary and Philosophical Society see note to Letter 70.

Letter 87.
CHARLES HUTTON to WILLIAM WINDHAM, 15 September 1802

London, British Library, Add. MS 37915, fol. 218–19: ALS.
Not addressed on the cover; the letter is preserved among a collection of the correspondence of William Windham.

Woolwich
Sepr. 15, 1802

Dear Sir

I am honoured by your note just received concerning Mr Hart's machine,[1] and in consequence have written to him by this post to appoint him a meeting here next Sunday to inspect it. — He had before merely announced himself to me by letter, without mentioning any other gentleman's or other recommendation, and being a total stranger, so much so indeed that I had never heard of his name, that I urged the fact of my numerous engagements as a reason to beg he would excuse me seeing it.

I believe the Club[2] will not meet again till the beginning of winter, about the time the Courts of Law begin to sit. — I understand the Bishop's 2d vol.[3] has made its appearance, though I have not yet seen it. It is employed it seems entirely on the Elements of Euclid, with some notes. — I have just received the 3d & 4th vols. of Montucla,[4] containing the history of the Sciences during the 18th Century: being two very large volumes indeed, it seems to me larger than was necessary for a mere history, beyond which character it is far extended, so as to enter largely into the critical & minuter parts of the sciences, which will set the book rather beyond the convenient and pleasant perusal of gentlemen in general. — I have lately contrived an instrument or method by which all the four conjugate hyperbolas are exhibited as cut at once from the same body by one plane section; which I shall be glad of an opportunity of shewing you some day when convenient. We have long had the method, you know, of getting the double section, or one opposite to another by two opposite cones, and then we have been indulged with the postulate that there may be two other opposite hyperbolas, in the same plane, having the same two axes as the two former, but only changed the one for the other, the transverse for the conjugate, & vice versa; but it has hitherto been a desideratum how to make those four sections by one cut of a solid. — I have lately had some trouble and vexation with a writer in the Monthly Review, who fell unmercifully & unjustly on my little book on Bridges.[5] This reviewer is a Mr Woodhouse, a tutor in Caius College Cambridge, who, from some private pique or motive, has for some years been in the practice of unjustly aspersing my publications in the Mo. Rev. His treatment became so glaring that I found it necessary to apply to Dr Griffiths, the proprietor of the Review, but not being allowed sufficient redress from that quarter, it became necessary to have recourse to some other chanel of stating my case: accordingly I got a paper inserted in the Monthly Magazine for August last,[6] by which I believe the matter has been set to rest.

I observe a letter published by Mr Cobbit, to Sir Jos. Banks, on the subject of his letter to the National Institute of France,[7] which I think one of the best compositions of the kind I have ever seen, and which seems to be exactly in the manner of a certain very respectable friend of ours. —

I have the hon^r to be, with great respect,
Sir
Your most faithful hble Servt.
Cha. Hutton. —

I understand it is probable the consideration of the new London bridge will be resumed in the ensuing Sessions of Parliament, & that other projects may be resorted to, instead of Mr Telford's plan, of a single arch, by adopting a plan of three or five arches.[8] In case of their preferring the plan of three arches, I have conceived an idea of that kind, which is quite new, & appears to possess more advantages of usefulness & beauty than any other, and which I should like to offer for their consideration, if I could know how.

The late consideration of a new London bridge, having given me an occasion to resume that subject, I have prepared a new treatise on it, on a much more extensive plan than the former.[9] And in consequence of the late disputes about the theory, I think of now adding to it the history of all the theories and other compositions that have been published of that kind.

Notes

For William Windham see note to Letter 78.

[1]According to MPHE, Naphtaly Hart

> took out a patent in 1801 for a gonimeter, an instrument 'laying down angles to minutes, whereby accurate triangles could be drawn for the graphic solution of problems of trigonometry, navigation, etc.'

Apart from this and another patent around the same time he seems to be unknown. See Letter 88 for Hutton's subsequent meeting with him.

[2]For the 'Friday' Club see note to Letter 20.

[3]Presumably Horsley, *Euclidis elementorum libri priores XII*, perhaps taking Horsley, *Elementary treatises* as his 'first volume'. For Samuel Horsley see note to Letter 5; he was at this time Bishop of Rochester.

[4]That is, Montucla, *Histoire*, of which the second two volumes were edited after Montucla's death by Jérôme Lalande (for whom see note to Letter 10).

[5]For Robert Woodhouse and his criticisms of Hutton, *Bridges* (1801) see Letter 85 and notes.

[6]That is, Hutton, 'To the Editor of the Monthly Review'.

[7]The 'respectable friend' was Samuel Horsley and the letter 'Misogallus' [Horsley], 'To Sir Joseph Banks', which appeared both in the *Weekly Register* and as a separate pamphlet during 1802. William Cobbett (1763–1835) was the publisher of the *Weekly Register*; for Joseph Banks see note to Letter 22.

[8]In March 1801 the Parliamentary Committee for the further improvement of the Port of London consulted Hutton about the single cast iron arch proposed by Thomas Telford (1757–1834) to replace London Bridge: he replied in some detail on 21 April. Hutton printed that (broadly positive) reply as Tract V in his *Tracts* (1812), vol. 1, pp. 127–44, and discussed the single arch (somewhat more sceptically), alongside the three- and five-arch designs also proposed by Telford and a three-arch design of Wilson, in the following tract on cast iron bridges: pp. 157–65. He set out his preference for a three-arch design on pp. 162–3; no more detailed account of the subject or of his 'quite new' idea seems to be extant. In the event, a five-arch design in stone by John Rennie (for whom see note to Letter 84) eventually opened in 1831.

[9]The reference was to Hutton, *Bridges*, reprinted in 1801 when it was criticized by Woodhouse (the 'late disputes about the theory' Hutton mentions: see Letter 85) and eventually included in its revised form in Hutton, *Tracts* (1812), vol. 1, pp. 1–115.

Letter 88.
Charles Hutton to William Windham, 19 September 1802

London, British Library, Add. MS 37915 fol. 220–21: ALS.

Not addressed on the cover; the letter is preserved among a collection of the correspondence of William Windham.

Woolwich Com^{n}, Sepr. 19, 1802.

Dear Sir,

Mr Hart has just been with me, with his very nice and curious instrument, with which I have been much pleased and amused.[1] In plane trigonometry, it resolves the cases of right-angled triangles very neatly to two places of figures, such as the cases of traverse sailing in Navigation, for the difference of latitude & departure; but not however so easily or expeditious as by the ordinary tables in books for that purpose. The instrument very readily and accurately lays down and draws any angles, & lines, parallel, perp^{r}. or oblique, &c, also for polygons, circles, & ellipses, & may be useful for cases of very great nicety & accuracy for certain curious purposes in Fortification, architecture, Dialling, Geography, &c. But I apprrehend [*sic*] that the appearance of complexness, that is about ⟨it,⟩ will always prevent it from being generally used in common concerns, & from superseding the use of common scale & compasses in such purposes. I have the hon^{r}. to be

Sir

Your most obed^{t} hble Servant

Cha. Hutton.

Note

For William Windham see note to Letter 78.

[1]See Letter 87.

Letter 89.
CHARLES HUTTON to JOSEPH BANKS, March 1803

London, Royal Society, CMO/8, p. 218: in a minute of the Council meeting of 31 March 1803.
The original letter is not extant.

A Note from Dr Hutton to the President was read signifying his intention to undertake the care of arranging and printing a new abridgement of the Philos: Transactions,[1] and trusting that the President and Council would please to countenance this undertaking. This communication was ordered to be entered on the minutes.[2]

Notes

For Joseph Banks see note to Letter 22.

[1]Hutton, Shaw, and Pearson, *The Philosophical Transactions ... abridged*. Although the natural historian George Shaw (1751–1813) and physician Richard Pearson (1764–1836; both ODNB) were named as joint editors on the title page, Hutton's obituarists confirmed what is implied here, that he was in overall charge of the project: see Gregory, 'Memoir', 216 and the obituary of Hutton in *The Edinburgh Annual Register* 16 (December 1823), pp. 328–31, at p. 330. Leybourn, 'Charles Hutton' went further, stating that (p. 193)

> The departments of pure and mixed mathematics, including the practice, as well as the theory, of mechanics, hydrodynamics, optics, astronomy, electricity, and magnetism, were undertaken by Dr. Hutton, in addition to the general editorship and correction of the press, of the whole. The greater portion of the biographical articles, as well as of the translations, were from the Doctor's pen[.]

The *Abridgement* was advertised in a prospectus dated 31 March 1803; Hutton's letter to Council evidently only just preceded this publicity, and must have come some considerable time after Hutton had in fact committed himself to the project.

[2]This minute—surely significantly—does not say whether the President and Council were pleased to countenance the *Abridgement*: in practice the Royal Society was not able to control the reprinting of material from the *Philosophical Transactions*: see Watts, 'We want no authors', *passim*. When the *Abridgement* began to appear in June 1803, however, the first volume bore a dedication to the President, Vice Presidents, Council and Fellows of the Royal Society 'With Permission'.

Letter 90.
CHARLES HUTTON to RICHARD ELSAM, 21 May 1803

Printed in Elsam, *Rural architecture*, 29.

WOOLWICH, May 21, 1803.

SIR,

I HAVE been honored with your note, requesting the solution of a problem, viz. to find the area, or superficial content of the segment of a circle, whose chord-line is 14 feet, and the versed sine, or height, is two feet eight inches.[1]

In answer to which I have to observe, that the area required is very nearly equal to 25 square feet, and $\frac{6}{10}$, or $\frac{3}{5}$ of a square foot, being more accurately 25.5969924 square feet, that is, 25 square feet, and the decimal .5969924, which decimal is true and correct in every figure of it. That is, on supposition that the versed sine is two feet eight inches, which I understand it to be, and not 2.8 or two feet eight-tenths.[2]

If it will be any satisfaction to you to see the whole process of the calculation, you will please to let me know, and I shall copy it out and send it.

I am, SIR,
Your obedient humble servant,
CHARLES HUTTON.

Mr. ELSAM.

Notes

Richard Elsam, architect, went on to publish *Hints for improving the condition of the peasantry ... by promoting comfort in their habitations* (1816) and *The practical builder's price-book* (1825), which ran to several editions up to 1847.

[1]Involved in a dispute with the district surveyor about some houses he was building, Elsam obtained the opinions of one architect and two mathematicians on this point, and then applied to Hutton for confirmation: Elsam, *Rural architecture*, 25–8.

[2]A footnote by Elsam indicated that this was the correct supposition.

Letter 91.
JOHN LESLIE to CHARLES HUTTON, 19 September 1804

Edinburgh, University Library, La. II. 425/30: ALS.

Cover lost; recipient inferred from library catalogue entry and internal evidence.

London 19 Septr 1804.
22 Poultry

Dea[r Sir,]

Allow me to ask you, whether in the enlarged plan of the Academy at Woolwich[1] there will be an opening for a professor of Natural Philosophy or Chemistry? And if such, what attendance would be required, & what emolument would be allowed? I have long had a wish to settle in the neighbourhood of the metropolis; but I find, that, without some considerable addition to my income, I could not live in a comfortable style. If the idea which I have now suggested & which has only within these few days entered my head could be realised, I should feel a very sensible pleasure in the prospect of enjoying your society.

I find that I cannot get away before Sunday, & if you can stea[*text lost*] laborious pursuits as to drop me [*text lost*] esteem it a farther mark of your friendship.

I beg to present my kind respects to Mrs Hutton & family, & am
Dear Sir
with very sincere esteem
Your obliged friend
John Leslie

Note

For John Leslie see note to Letter 72.

[1]The Royal Military Academy was reorganized several times during Hutton's time there, and the mathematical staff in particular was enlarged in the winter of 1802–3. Hutton himself had lectured on natural philosophy from May 1799, but no professor in that subject was appointed. However, the post of lecturer on chemistry was at this time being filled by John MacCulloch (1773–1835: ODNB) on a temporary basis until his appointment was confirmed in October, and Leslie may have believed that a vacancy existed: Jones, *Records of the RMA*, iii–iv, 49, 55.

Letter 92.
CHARLES HUTTON to FRANCIS WINGRAVE, 1805

Newcastle, Literary and Philosophical Society, N925/1 (copy of Bruce, *Memoir*): ALS tipped in at end.

Recipient identified in endorsement. Year deduced from internal evidence.

The enclosures mentioned in the letter are not extant.

Sir

You will perceive that in the inclosed sheet (F), there occur both a wooden cut & a copper plate.[1]

Be so good also as complete the direction of the inclosed letter, ⟨to Bp. Horsley,⟩[2] by adding his (town) residence, if you think it necessary, (as I do not know it), & put the letter in the post for him, or otherwise let him have it.

Your hble Serv[t]
CH.

Mr Wingrave

Notes

Francis Wingrave, bookseller, had in 1789 taken over the firm of John Nourse (a mathematical specialist, among other things, and one of the publishers of Hutton, *Tables*): ODNB *s.v.* 'Nourse, John (bap. 1705, d. 1780), bookseller'. He published Hutton's edition of Thomas Simpson's *Select Exercises* (1792) and his edition of Benjamin Robins' *New Principles of Gunnery* (1805). This letter can be dated from the reference to the printing of the latter; see note 1.

[1]Hutton's edition of Benjamin Robins' *New Principles of Gunnery* (1805), which Francis Wingrave published, indeed contains both a woodcut and a copperplate within sheet F, and on this basis the letter is dated to 1805.

[2]For Samuel Horsley see note to Letter 5.

Letter 93. M. Courtois to Charles Blacker Vignoles, 18 January 1805

Portsmouth, History Centre, Vignoles Papers, 1072A/App. 14: ALS (our source). Printed in Vignoles, *Infant Ensign* as Appendix 14.

Paris le 18. Janvier 1805.

Dans le désordre du siége, lors de la reprise par les français de la Ville Pointe à pitre guadeloupe, en 1794. Je donnai l'asile à M. Vignolles officier au 43e. Régiment et à son épouse qui tenoit un enfant dans ses bras: c'étoit vous, Charles, vous n'aviez pas un an.[1]

La fiévre jaune éxerçoit ses ravages, votre Pere & votre mere en furent victimes: mes soins n'ayant pu les sauver: Je leur fis donner une décente sépulture.

Vous fûtes également atteint de cette cruelle maladie: Le lait altéré de votre mere en avoit porté le germe dans votre sang: une nouvelle nourrice vous conserva à votre famille.

Dix[a] mois s'étoient écoulés quand le Capitaine d'artillerie Hutton, votre oncle, vînt en parlementaire, vous réclamer.[2]

Quelle joie il eut de vous trouver au milieu des horreurs où nous étions livrés! Je lui remis, avec votre personne, le prétieux écrit que sa sœur, votre mere, traçait dans ses derniers momens, lorsque s'entretenant du chagrin que la nouvelle de sa mort causeroit à son Pere qui s'étoit opposé à son départ,[3] et voulant lui faire ses adieux: la plume tomba de ses mains.

Une note à La suite[4] fait mention de l'époque où vous avez perdu les auteurs de vos Jours, et porte l'expression des vœux que je formois, pour que vous vous rendissiez heureusement au sein d'une famille dont vous deviez être la consolation.

Vous partîtes. Je n'ai plus entendu parler de vous, mais si vous existez: vous gardez souvenir de moi ... [*sic*]

J'ai un ami particulier prisonnier en Angleterre: Il se nomme Lamoziniere et je le suppose à Civertown.[5] Parti des Indes occidentales en pleine paix: Il revenoit sur le navire le Bonaventure avec toute sa fortune, et il a été pris à la vue des côtes de france. Il est ruiné. Mais ce seroit un bonheur pour lui et pour moi qu'il fut renvoyé sur sa parole.

Les circonstances s'étant opposé [*sic*] au succès des démarches qui ont été faites jusqu'à présent: Je commençois à désespérer, lorsque je me suis rappellé de La famille du célébre Docteur Hutton, et j'ai pensé qu'elle me serviroit en cette occasion, si elle pouvoit.

Je ne doute point que si cette lettre vous parvient: vous ou vos parens ne fassiez les sollicitations nécessaires pour obtenir la liberté de mon ami; et si vous ne réussissez pas: Je n'en serai pas moins persuadé que la chose n'aura point dépendu d'eux ni de vous.

Adieu, Charles, Je souhaite que vous soyez heureux,

Courtois.

M[r]. Charles Vignolles
En son absence
à La Famille du Docteur Hutton
à Londres.

---

[a] Altered in pencil, in another hand, to: *Deux*.

[*Pencil addition*]: Lamozinière, as I have heard my father say was released through Dr Hutton's influence.

Hutton Vignoles.[6]

Notes

For M. Courtois see note to Letter 66; for Charles Blacker Vignoles see note to Letter 61. Since the latter was only 11 when this letter was written, there seems no doubt that it was intended for the eyes of Charles Hutton.

[1]See Letter 66 for another account of Charles Henry Vignoles, his wife Camilla, and his son Charles Blacker Vignoles in Guadeloupe.

[2]See notes to Letters 66 and 71 for Henry Hutton's rescue of Charles Blacker Vignoles. The deaths of Charles Henry and Camilla occurred in June and the appearance of Henry Hutton in December (see Letter 66): 'Dix' was therefore a mistake not for 'Deux' but for 'Six'.

[3]It is no surprise that Hutton was opposed to Camilla's departure for that notoriously unhealthy war zone, the West Indies, but this is our only direct evidence for the fact. It should perhaps be related to his silence about her death in Letter 67 and his brief, even casual reference to the event in Letter 71, and possibly also to Courtois' report in Letter 66 of Camilla's final anxiety with regard to her family ('me pria de leur témoigner de sa part tout le regrêt qu'elle avoit de mourir loin d'eux').

[4]That is, Letter 66.

[5]Presumably a mistake for Tiverton in Devon, a 'parole town', where officers on parole were permitted to reside, in the period 1803–13: Abell, *Prisoners of War*, 298–300. I have not traced Lamoziniere.

[6]Hutton Vignoles (1824–89), engineer, was the second son of Charles Blacker Vignoles: obituary in *Minutes of the Proceedings of the Institution of Civil Engineers* 100 (1890), pp. 406–8.

Letter 94. JUDITH MILBANKE to CHARLES HUTTON, 10 February 1805

Printed in Hutton, *Recreations* (1814), vol. 4, pp. 223–7.

Sender identified from Letter 96; also identified in 'The divining rod', *Newcastle Magazine* 2 (January 1823), 48–50 and Hutton, *Recreations* (1840), 712.

Feb. 10, 1805.

Sir,

Having lately read in your translation of Montucla's Recreations the article respecting the *Divining Rod*,[1] I am tempted to trouble you with an account of the *Real Powers* it may possess. From the character I have universally heard of your candour, I am inclined to think you will place confidence in what I shall relate; it may also afford you some amusement, and should you, from reflecting on the *Facts*, discover the *Cause*, it would be a great gratification to me to be informed of your discovery.

I have perused the history of La Baguette Divinatoire;[2] in that book much superstition and fable are blended with the simple truth: natural philosophy had made but little progress at the period that book was written, and they joined together things which had no other connection than from erroneous ideas: even in these more enlightened days the absurdity of animal magnetism, founded on some allusion to the real magnet, has found partizans and believers; yet that folly cannot shake the conviction of the influence of the magnet, though not yet I believe satisfactorily accounted for.

In the year 1772 (I was then 19) I passed six months at Aix in Provence: I there heard the popular story of one of the fountains in that city having been discovered some generations before, by a boy who always expressed an aversion from passing one particular spot, crying out *there was water*. This was held by myself, and the family I was with, in utter contempt, and believed as much as the tradition of this county, that St. Dunstan's head rolled to the spot where Durham abbey was to be built and dedicated to him.

In the course of the spring the family went to pass a week at the chateau D'Ansouis, situated a few miles to the north of the Durance, a tract of country very mountainous, and where water is ill supplied. We found the marquis D'Ansouis busied in erecting what might be termed a miniature aqueduct, to convey a spring the distance of half a league, or nearly that distance, to his chateau, which spring he asserted had been discovered by a peasant, who made the discovery of water his occupation in that country, and maintained himself by it, and was known by the appellation of *l'homme à la Baguette:* his real name has escaped my recollection; but you will find some account of him in one of the Annual Registers near the period I have named, I believe that of 1772[3]; but not at present having the books to refer to, I cannot ascertain the year.

This account was received with unbelief, almost amounting to derision. The marquis, piqued at being discredited, sent for the man, and requested we would witness the experiment. A large party of French and English accordingly attended. The man was quite a peasant in manners and appearance: he produced some twigs cut from a hazel of different sizes and strength, only they were the forked branches, and hazel was preferred as forking more equally than most other trees; but it is not requisite that the angle should be of any particular number of degrees. He held the ends of the twigs between each fore fingers and thumbs, with the vertex pointing downwards. Standing where there was no water, the baguette remained motionless; walking gradually to the spot where the spring was *under ground*, the twig was sensibly affected; and as he approached the spot, began to *turn round;* that is, the vertex raised itself, and turned towards his body, and continued to turn till the point was vertical; it then again descended outwards, and continued to turn describing a circle, so long as he remained standing over the spring, or till one or both the branches were broken by the twisting, the ends being firmly grasped by the fingers and thumbs, and the hands kept stationary, so that the rotary motion must in course twist them. After seeing him do this repeatedly, the whole party tried the baguette in succession, but without effect. I chanced to be the last; no sooner did I hold the twig as directed, than it began to move as with him, which startled me so much that I dropt it, and felt considerably agitated. I was however induced to resume the experiment, and found the effect perfect. I was then told it was no very unusual thing, many having that faculty, which, from what has since come to my knowledge, I have reason to believe is true. On my return to England I forbore to let this faculty (or whatever you may term it) be known, fearing to become the topic of conversation or discussion. But, two years afterwards, being on a visit to a nobleman's house in Huntingdonshire,[4] and his lady lamenting that she was disappointed of building a dairy-house in a spot she particularly wished, because there was *no water* to be found, a supply she looked on as essential; under *promise of secrecy* I told her I would endeavour to find a spring. I accordingly procured some hazel twigs, and, in the presence of herself and husband, walked over the ground proposed, till the twig turned with *considerable force.* A stake was immediately driven into the ground to mark the spot, which was not very distant from where they had before sunk. They then took me to another, and distant building in the park, and desired me to try there: I found the baguette turn *very strong*, so that it soon twisted and broke: the gentleman persisted that there was no water there, unless at a great depth, the foundation being very deep (a considerable stone cellar) and that no water appeared when they dug for it. I could only reply, that I have no more than from the baguette turning, and that I had too little experience of its powers or certainty to answer for the truth of its indication. He then acknowledged that when that building was erected, they were obliged to drive piles for the whole foundation, as they met with nothing but a quicksand. This induced him to dig in the spot I first directed; they met with a very fluent spring; the dairy was built, and is at this time supplied by it. I could give a long detail of other trials I have made, all of which have been convincing of the truth, but they would be tedious. For some years past I have been indifferent about its becoming known, and have consequently been frequently requested to show the experiment, which has often been to persons in high estimation for understanding and knowledge, and I believe they have *all been convinced.* Three people I have met with who have, on trying, found themselves possessed of the same faculty. I shall only add one more particular incident: having once shown it to a party, we returned into the house to a room on the ground floor; I was again asked to show *how I held the twig;* taking one in my hand I found it turned immediately; on which an old lady, mother to the gentleman of the house, said *that room* was formed out of an old cloister, in which cloister was a *well*, simply boarded over when they made the room.

L'homme à la baguette from experience, could with tolerable accuracy tell the depth at which the springs were, and their volume, from the force with which the baguette turns; I can only give a rough guess. In *strong frost* I think its powers not so great; on a bridge or in a boat it has *no effect;* the water must be *under ground* to affect the baguette, and running through wooden pipes acts the same as a spring; even being *arched* over does not prevent it, as I felt a spring broke out in a cellar, standing in a stone passage over the arch. I can neither make the baguette turn where there is *no water*, nor prevent it from turning where there *is* any, and I am perfectly ignorant of *the cause why it turns.*

The only sensations I am sensible of, is an emotion similar to that felt on being startled by sudden noise, or surprise of any kind. If, sir, you can throw any light on this fact, and account for it, I hope you will indulge me by communicating your conjectures. A letter under cover to —— near ——,[5] directed to x. y. z. will by him be forwarded to me. If you should think it worth your trouble to make any queries on the subject, I will with pleasure reply to them.

I am, Sir,

your humble servant, and admirer of your talents,

x. y. z.

Notes

Judith Milbanke, born as the Hon. Judith Noel (1751–1822), was the daughter of Edward, first Viscount Wentworth; in 1777 she married Sir Ralph Noel, baronet. From 1815 the family name changed to Noel by royal licence. Information from www.thepeerage.com/p2747.htm; see also Letters 96, 97, 99, and 107.

Judith's daughter Anne Isabella (1792–1860) married Lord Byron the poet in 1815; their daughter Augusta Ada became a patron of Charles Babbage (for whom see Letter 114).

[1]Hutton, *Recreations*, vol. 4, pp. 259–67, where the text ridiculed what it called an 'illusion, or philosophical quackery' (pp. 259–60). I have suggested in Wardhaugh, *Gunpowder and Geometry*, that Hutton exercised no more than a supervisory role with respect to the book.

[2]Le Lorrain de Vallemont, *La baguette divinatoire*, also mentioned in the *Recreations* passage.

[3]It is at p. 96, pt. 2, of that year. Editor. [Note in source.]

[4]In Letter 97 this house was identified as Kimbolton, the seat of the Dukes of Manchester.

[5]It appears from Hutton's reply, Letter 96, that Milbanke gave the name and address of her husband Sir Ralph Milbanke here; they were suppressed when the letter was printed in 1814.

Letter 95. CHARLES HUTTON to WILLIAM FETTES, 19 February 1805

Edinburgh, City Archives, SL1/1/142: Town Council Minutes vol. 142, pp. 139–41: copy. Recipient inferred from context.

Royal Military Acad: Woolich
February 19th. 1805

My Lord

I entreat your Lordship's forgiveness for the liberty thus taken by an unknown Individual in addressing your Lordship as head of the Magistracy of Edinburgh, Patrons of the University of that City, concerning the Mathematical Professorship now vacant, in favor of M^{r}. John Leslie a Candidate for that Office. This I am emboldened to do both by the known protection of literary merit manifested by the Magistracy of Edinburgh, and by the firm belief that I am endeavouring to render an essential service to the interest of the University, by thus recommending a most ingenious & valuable man who has for many years ⟨been⟩ intimatly known to me as a profound Mathematician and Philosopher — Should M^{r}. Leslie become the choice of the Magistrates, I have no hesitation, My Lord, from my personal knowledge of his Talents, and from my experience gained in the course of a long life spent in the same profession, to declare my firm belief of M^{r}. Leslies filling the Chair with honour to himself and benefit to the City & University —

I have the honour to be with great respect

My Lord

Your Lordships faithful

humble Servant

Chas. Hutton

Note

Sir William Fettes, baronet (1750–1836) was Lord Provost of Edinburgh at this time: ODNB. As Morrell, 'The Leslie Affair', 77 narrates, the candidacy of John Leslie (see note to Letter 72) for the mathematical chair at Edinburgh was controversial and

> [i]n order to justify its election of Leslie, the Edinburgh town council unanimously supported the lord provost's highly unusual suggestion that testimonials and letters supporting Leslie should be recorded. They occupy Town Council Minutes, cxlii, 139–200, 20 Mar. 1805.

Leslie's supporters, among whom Hutton was recorded first, included John Robison, Francis Maseres, Nevil Maskelyne, and John Playfair.

Letter 96.
CHARLES HUTTON to JUDITH MILBANKE, 15 April 1805

Oxford, Bodleian Library, Dep. Lovelace Byron, box 13, fols. 93–4: an 1803 prospectus for Hutton, Shaw, and Pearson, *The Philosophical Transactions ... abridged* with ALS attached.

Woolwich, April 15, 1805.

Sir

I was duly favoured with your letter of Feb. 10,[1] giving some curious and ingenious accounts of the Divining Rod, and requesting an opinion on the cause, &c. The accounts you give, Sir, are very curious and circumstantial. Before this, Sir, notwithstanding the general notices, both ancient and modern, & what is even contained in the Recreations,[2] I must fairly acknowledge that I had no belief in the fact, but considered the several accounts as founded in artifice or deception. But from the accounts in your own instance, and related with such appearance of candour and sincerity, I feel much staggered in my opinion, and after which there seems to be nothing wanting to my full conviction but the knowledge of the name &c of the person who has favoured me with the interesting particulars. If it be not improper then, may I request you will have the goodness to inform me on that head, as also where it was in Huntingdonshire &c that you experienced the said effects. With your leave also I would notice the circumstances in the next edition that may be printed of the Recreations &c. also in a note to any paper I may meet with [in] the Philos. Trans. in the Abridgment to which the annexed printed paper relates, though I do not yet know if there is any paper on this subject in those Transactions.[3] – But as to the cause of this curious phenomenon, which you request, I must ing⟨en⟩uously acknowledge I have not the least knowledge, either as to the means or the probability in any way whatever. – I have tried the rods myself, as well as several of my acquaintances, but without effect, viz. over a well, quite covered over, & having a pump in it. I would enquire if the rod is to be cut off close to vertex of the angle, or whether ⟨there⟩ is to be a part of the thicker single stem beyond the vertex. I would only add that the Annual Register you allude to without date, is that for the year 1772, & is in p. 96 pt 2. And am, Sir,

Your obliged hble Servant,
Cha. Hutton.

For X. Y. Z.
To the Care of
Sir Ralph Milbanke Bar^{t}.
Seaham.
near Stockton on Tees

Notes

For Judith Milbanke see note to Letter 94; see also Letters 97, 99, and 107. At this stage Hutton knew her only as 'x.y.z.'; he may have assumed that he was in fact writing to Sir Ralph Milbanke (her husband), whose address she had given.

[1]That is, Letter 94.

[2]Hutton, *Recreations* (1803), vol. 4, pp. 259–67, where various reports connected with the divining rod were (with considerable scepticism) narrated.

[3]There is no entry for 'divining rod' in the index to Hutton, Shaw, and Pearson, *The Philosophical Transactions . . . abridged*.

Letter 97.
JUDITH MILBANKE to CHARLES HUTTON, 30 April 1805

Printed in Hutton, *Recreations* (1814), vol. 4, pp. 228–9.
Sender identified from Letter 96; also identified in 'The divining rod', *Newcastle Magazine* 2 (January 1823), 48–50 and Hutton, *Recreations* (1840), 712.

_____[1]

April 30, 1805.

Sir,

I have received your letter directed for x. y. z.[2] in reply to one bearing that signature of Feb. 10, 1805.

The candid manner in which you have considered the narrative I then sent demands an acknowledgment. I withheld my name, more from a wish that you might openly state any doubts you might entertain, than from any wish to conceal it, as otherwise you might have felt some delicacy in expressing disbelief. I lament you can throw no light on this extraordinary circumstance,[3] which has ever strongly excited my curiosity as to the cause of it, but hitherto I have met with none who have gone beyond a vague conjecture. A very sensible and well informed physician *imagined* it might be occasioned by some singular effect of electricity on my frame, but could not satisfy himself of the certainty of his conjecture. I shall most probably be in London next winter, and will (if you wish it) afford you an opportunity of making your own observations on this curious fact, which you may then give any account of that you please, either in the next edition of your Montucla, or any other publication.

The nobleman's house I named in my letter was Kimbolton, and I make no doubt that the present dowager duchess of Manchester well remembers the experiment and its success.

I could refer you to many of the first characters in the kingdom who have seen it tried at different times and different places, but am inclined to be of opinion that you will not suspect me of a wish to mislead you and expose myself by a ridiculous falsehood.

I enclose you a *baguette*, exactly as I use it, but a larger one (inconvenient to send in a letter) shows the power more fully, as so small a one almost immediately breaks from the twisting of it, near the finger and thumb. I shall only add that many trials since the year 1772 have *convinced me*, that *the fact is certain*, for I myself was slow to credit it. Should you wish to write again, do me the favour to direct to the Honble. —— at ——.

I have the honour to be, Sir,

your obedient humble servant,

—— ——

I generally use a baguette about six inches from the vertex to the ends of the twigs where they are cut off.

Notes

For Judith Milbanke see note to Letter 94; see also Letters 96, 99, and 107.

[1]Although Hutton omitted the name when printing the letter in the second edition of the *Recreations*, it is evident that the author here and below named herself as the Hon. Judith Milbanke, presumably at Seaham near Stockton-on-Tees.

[2]That is, Letter 96.

[3]That is, the behaviour of the divining rod: see Letters 94 and 96.

Letter 98.
CHARLES HUTTON to ANDREW MACKAY, 5 April 1806

Cambridge, University Library, RGO 14/44, fol. 120: copy.
The sheet also contains copies of letters from Andrew Mackay and George Gilpin on the same subject.

Woolwhich 5th april 1806.

Dear Sir

the bearer, Viscount de Vaux, having a scheme for the Longitude,[1] which appears to be ingenious, I take the liberty of recommending it to your attention, being a Subject in which you are more intimately conversant than any other persons, and I am

Dear Sir &c (signed) Ch: Hutton.

to Dr MacKay Mathematical examiner Trinity house &ca

Notes

For Andrew Mackay see note to Letter 82.

[1]Charles Grant, Vicomte de Vaux in the peerage of France (1749–*c.* 1818), lived in Britain from 1790 and supported himself by writing on various subjects. Information from quod.lib.umich.edu/c/clementsmss/umich-wcl-M-305vau?view=text; also Weatherford, 'The Vicomte De Vaux'. He published his longitude scheme as Grant, *Longitude*; it involved dead-reckoning using an improved log. Although Mackay gave it his support it was rejected by the Board of Longitude: see the review in the *Monthly Review* (September 1810), pp. 83–7.

Letter 99.
JUDITH MILBANKE to CHARLES HUTTON, 23 May 1805

Printed in Hutton, *Recreations* (1814), vol. 4, p. 229.

May 23d.

Lady ——'s[1] compliments to Doctor Hutton, and proposes to be at Woolwich on Friday the 30th inst. at eleven in the forenoon, if that day and hour will be convenient to Doctor Hutton. Lady —— will then have the pleasure of showing him the effect of the *Divining Rod*, and requests he will provide some branches of hazel, which she will cut into the proper forms when she arrives.

Doctor Hutton will have the goodness to answer Lady ——, to inform her if he shall be at leisure on Friday to receive her at his house, for the purpose of the experiment.[2]

Notes

For Judith Milbanke see note to Letter 94; see also Letters 96, 97, and 107.

[1]Lady Milbanke had now identified herself to Hutton (see Letter 97), but her name was concealed when this series of letters was printed in 1814.

[2]No reply is extant, but Hutton did indeed receive Lady Milbanke as requested, with her family, at his house at Woolwich Common. In the presence of Hutton's family and some friends she performed the divining-rod experiment several times, illustrating to Hutton's satisfaction that the rod moved as described in Letter 94 at places where there was water below the surface. See Hutton, *Recreations* (1814), vol. 4, pp. 229–30.

Letter 100.
GILBERT AUSTIN to CHARLES HUTTON, 13 August 1806

Portsmouth, History Centre, Vignoles Papers, 1072A/O1: ALS.

Augt. 13. 1806
Woodville
Leixlip
Ireland

Dear Sir

By this nights Post I have written to my booksellers Messrs. Cadell & Davies in the Strand requesting that they would on my part present you with a copy of a work which I have just published named Chironomia or a Treatise on Rhetorical delivery.[1] You may recollect that I was about publishing this book last summer when M^{rs}. Austin my little nephew Bushe & I had the pleasure of seeing you at Woolwich. At which time you were so good to say you would accept of a Copy as a mark of my respect. I hope my dear little Cousin Vignolles[2] is well & proceeding in his studies so as to merit your approbation. Little Bushe conceived a strong & sudden friendship for him, I hope it may be cultivated in more advanced life & become permanent. They are both such boys as may be expected to distinguish themselves in life. He desires to be kindly remembred to little Vignolles as does also M^{rs}. Austin — as to myself I am strongly interested in the happiness of so near a relative & hope he will consider me to be so. We beg also our best Comp^{ts}. to M^{rs}. Hutton & the Ladies.[3] I have not yet been so fortunate as to get introduced to Col^{l}. Hutton[4] but should peace ensue I hope his engagements will be less strict & that he will have it in his power to visit Dublin. In such an event I beg he may have the goodness to consider my family as anxious to be known to him. I am D^{r}. Sir with the highest esteem & respect yours

Gilb^{t} Austin

Should it be convenient to you Messrs. Cadell & Davies would give the book to your order.

Doctor Hutton
Royal Military Academy
Woolwich

Notes

The Rev. Gilbert Austin, M.A. (d. ?1835) was the son of one of Charles Hutton's sisters; he evidently visited Hutton in the summer of 1805 with his wife and a nephew ('Bushe'). In 1830 Charles Blacker Vignoles visited him in Dublin '& accompanied him on a visit to Mrs Chamley of Bullcamp another Cousin; both of these relations are Children of my Grandfather's Sisters': British Library, Add. MS 35071, Vignoles Journals 1830–3; entry for 1 January 1830. Unfortunately nothing more is known about the relationship. Chaplain of the Magdalen asylum in Dublin in the 1790s, Austin also published sermons and appears to have supplied words for an opera entitled 'Psyche' by J.A. Stevenson: information from ESTC.

[1]Austin, *Chironomia*.

[2]For Charles Blacker Vignoles see note to Letter 61.

[3]'The Ladies' included Isabella Hutton; Austin's use of the plural may indicate that her surviving sister Eleanor had by this time been widowed and returned to the Hutton household: see notes to Letter 16.

[4]For Henry Hutton see note to Letter 12.

Letter 101. CHARLES HUTTON to MESSRS KENT, ?1807

Oxford, Bodleian Library, MS Montagu d. 7, fol. 562: ALS.
Year deduced from internal evidence.

34 Bedford Row
Monday Novr. 16.

Doct^{r}. Hutton will be obliged to Messrs Kent, (the other room & the Stairs being now ready) to send the Carpets and lay them down, on Monday Morning next, the 23d instant.

Note

Messrs Kent were evidently a firm supplying carpets in London.

Between Hutton's move to Bedford Row in 1807 and his death in 1823, 16 November was a Monday in 1807, 1812, and 1818. I place this letter in 1807 on the assumption that the installation of carpets is more likely to have taken place soon after the move.

Letter 102.
CHARLES HUTTON to FRANCIS BAILY, 13 July 1808

London, Senate House Library, [DeM] L.4 [Waring] (copy of Waring, *Algebraic Quantities*), letter tipped in at end.
In the hand of Isabella Hutton; address and possibly signature in that of Charles Hutton.
Address from cover is pasted onto main sheet.
Year inferred from date of Letter 103.

34 Bedford Row July 13

Sir,

Indisposition for some days has delayed the expression of my thanks for your very neat and ingenious treatise on Interest and Annuities.[1]

With regard to the Pamphlet published at Cambridge, I wish it was in my power to give you more information than I can at present. That whole article, concerning Dr Waring's improvements, in my Dictionary from p. 717 to 726, was communicated by Dr W. himself.[2] And as he mentioned the Pamphlet in question, I had no doubt that it was a former publication of his, tho' I had never seen or heard of it before. And it is no less singular that the Pamphlet has never been mentioned in any one of ⟨the accounts of⟩ D^r^. W's life and writings. In this state of uncertainty, since the receipt of your note, it occured to me that the Pamphlet might be one of those 4 pamphlets that passed between Dr W. and Dr Powell, at the time of Dr W's becoming a candidate for the professorship.[3] But on looking into these pamphlets, of which I have copies, I could find nothing of the kind of the matters referred to in the Pamphlet. Fortunately however I have met with another referment to it, by which you may receive some satisfaction, viz. in Mr Wood's Elements of Algebra, the 3^d^ ed. published at Cambridge in 1801,[4] the end of part 3, pag. 275, Dr Waring is mentioned as a writer on life annuities. It is therefore very probable that Mr Wood or Mr Vince or many other gentlemen of that University can procure you the Pamphlet. I shall be glad to know the success of your further enquiries after it and I am, with great respect, Sir,

Your obed^{t} Servt.
Chas. Hutton

Mr. Francis Baily
N° 3 Verulam Buildings
Grays Inn

Notes

Francis Baily (1774–1844), stockbroker, had a career in business and published a definitive set of tables on leases and annuities; in retirement he would become a noted astronomer: ODNB, DSB, MPHE; see also Letters 103 and 125.

[1]That is, Baily, *Interest and annuities*.

[2]For Edward Waring see Letter 76. The 'addenda' printed at the end of Hutton, *Dictionary* (1795), vol. 2 included at pp. 716–26 a passage on 'Affected equations' containing a long discussion (pp. 717–26) 'Of Dr. Waring's Meditationes Algebraicæ' (the author of the discussion was not named). Within it, at p. 726, mention was made of 'a Pamphlet published at Cambridge', without further explanation.

[3]Namely Powell, *Observations*, and Powell, *Defence*, and the replies Waring, *Letter* and Wilson, *Vindication*. William Samuel Powell (1717–75: ODNB) was at the time a Fellow of St. John's College, Cambridge (he was subsequently master of the college); John Wilson (1741–93: ODNB, DSB), a student of Waring's at Peterhouse, would go on to be Senior Wrangler in 1761 and subsequently to a legal career. Waring was a candidate for the Lucasian Professorship in 1760 and Powell's (unsuccessful) attack was motivated by his support for another candidate: ODNB.

[4]Wood, *Algebra*, of which the third edition was indeed dated at Cambridge in 1801. James Wood (1760–1839), at this time a college tutor, was subsequently Master of St John's College, Cambridge and Dean of Ely: ODNB. Samuel Vince (1749–1821), mentioned below, was Plumian Professor of Astronomy and Experimental Philosophy at Cambridge from 1796: ODNB.

Letter 103. Francis Baily to Charles Hutton, 14 September 1808

London, Senate House Library, [DeM] L.4 [Waring] (copy of Waring, *Algebraic Quantities*), draft tipped in at end.

Sept. 14. 1808

Sir,

Agreeably to your request I now make known to you the result of my enquiries respecting Dr. Waring's treatise on Annuities[a].[1]

In the year 1792 he published a small pamphlet in octavo entitld "On the principles of translating &c; & which is (as I imagined) the very work alluded to in your Mat: Dict: Vol. 2. p. 726. It consists of 59 pages, the last 32 of which are appropriated to the subject of Annuities[b]

I am at a loss to conceive what object he had in view in publishing this part of his work; for certainly he has[c] thrown no new light on the subject. His problems ⟨(if a string of detached observations are worthy of that name⟩ are quite elementary[d] & his loose & illogical[e] method of treating[f] them add neither grace nor dignity to the work. ⟨The very title of the book betrays the inaccuracy of his style⟩[2] The only points which may be considered as interesting are

1°. A new formula for expressing the value of Prob III in Mr. Simpson's treatise on Annuities;[3] viz $S - m.S' + \frac{m.(m+1)}{2} \times S'' - \frac{m.m+1}{2}.\frac{m+3}{3}S''' + \&c$ where[g] S, S', S'', S''' &c denote respectively the sum of the present values of an annuity on every $m, (m+1), (m+2), (m+3)$ &c joint lives. This c...y is more elegant than the one in p. 26 of Mr. Simpson's work above mentioned

2°. Observations on the method of[h] determining the value of assurances when the sum is to be received immediately on the failing of the given lives: but, since in all practical cases,[i] such sum is not due till the end of the year in which the lives[j] fail, I look upon this as matter of curiosity[k] more than of real utility: particularly as he points out no rules for determining such value, neither do I see how the principles he has laid down could be applied with good effect to an assurance on the whole life;[l]

[*Text ends.*]

---

[a] *altered from*: public work on the subject of Ann.

[b] *altered from*: in question.

[c] *deleted*: added nothing new to our information.

[d] *deleted*: and after the be.

[e] *altered from*: obscure.

[f] *deleted*: the subject.

[g] *deleted*: $S =$ [*altered from: denotes*] the sum of the ⟨present⟩ values of an annuity on every m joint lives, S' ⟨=⟩ the sum of the [*altered from: value*].

[h] *altered from*: Proof for.

[i] *deleted*: the.

[j] *altered from*: life.

[k] *altered from*: a question of speculation.

[l] *deleted*: although it must be used for one.

Notes

For Francis Baily see note to Letter 102. Although he presumably sent some response to Letter 102, it is not clear how or whether it related to this incomplete draft.

[1]See Letter 102. Evidently Baily wrote to James Wood as Hutton suggested there; Wood replied on 5 September in a letter now tipped into the same volume as this one. It described Waring, *Algebraic Quantities*, and offered the loan of his copy if necessary.

[2]The full title of Waring's book was 'On the principles of translating algebraic quantities into probable relations and annuities, &c.' Baily evidently objected to the construction 'principles of translating'.

[3]That is, Simpson, *Annuities and reversions*, 25–6; the problem was 'To find the value of an annuity granted upon any number, n, of lives, A, B, C, &c. but so as to continue only as long as a given number, m, of them are in being.'

Letter 104. CHARLES HUTTON to JOHN BRUCE, 3 August 1809

Extract printed in Bruce, *Memoir*, 25–6.

London, Bedford Row, Aug. 3, 1809.

I am obliged to you for the printed pamphlets which I have received,[1] and beg you will have the goodness to remember and favour me, from time to time, with every thing that may be printed by and concerning the Library [*sic*] and Philosophical Society. And I rejoice at the fair prospect that is now before the Society, and that the philosophical lectures are likely to proceed with so promising effect under so able an agent as Mr. Turner.[2]

I am, dear Sir,
Your faithful servant,
CHAS. HUTTON.

To Mr. John Bruce.

Notes

John Bruce (1775–1834) was with his brother Edward (for whom see note to Letter 86) a schoolmaster in Newcastle from 1793; both were contributors to the *Ladies' Diary*, and John was personally acquainted with Hutton from at least 1804. He was the author of the most detailed life of Hutton to appear in the nineteenth century: Bruce, *Memoir*. Bruce, *John Collingwood Bruce*, 1, 3–4; Welford, *Men of Mark*, vol. 1, pp. 408–15; also Letters 106, 109, 115, 112, 113, 116, 118, 117, and 129.

[1]Presumably Bruce had sent Hutton some selection of matter relating to the Newcastle Literary and Philosophical Society (for which see note to Letter 70). Since 1796 the society had produced annual reports and lists of new members, as well as in 1801 a printed list of rules and catalogue of books and in 1807 a 'historical sketch' of its activities.

[2]William Turner (1761–1859), Unitarian minister and founder of the Newcastle Literary and Philosophical Society, was the society's lecturer on natural and experimental philosophy from 1802 to 1833: ODNB; Welford, *Men of Mark*, vol. 3, pp. 541–5.

Letter 105.
CHARLES HUTTON to WILLIAM WINDHAM, 15 December 1809

London, British Library, Add. MS 37910, fol. 85–6: ALS.
Not addressed on the cover; the letter is preserved among a collection of the correspondence of William Windham.

Bedford Row, Decr. 15, 1809.

Dear Sir,

Perceiving your name as a governor of the Charter House, may I be pardoned for soliciting an appointment, as a pupil on that establishment, for my grandson, Charles Henry Hutton,[1] who is turned of 9 years of age, and is a very hopeful youth, with whose maintenance and education I have charged myself. You would know, Sir, by a former card of mine, that that [*sic*] I resigned my situation in the Woolwich Academy above two years since.[2] — Not finding you at home when I have called, I take the liberty of leaving this note. And have the honour to be,

Sir
Your obliged humble Servant
Charles Hutton.

The Right Honb. Wm Windham.

Notes

For William Windham see note to Letter 78.

[1]Charles Henry Hutton, (1800–15), only child of Henry Hutton and his first wife Mary Anne Hodgson. He did not in the event enter the Charter House School (although, confusingly, another Charles Henry Hutton and another Henry Hutton did so around this

time). He later became a cadet at the Royal Military Academy, but he died aged 15, probably of consumption. Information from Portsmouth, History Centre, Vignoles Papers, Letters 38, 72, 75, 76/1, and 80; 'Charterhouse Register 1769–1872' (accessed via www.charterhouse.org.uk/archives), pp. 204–5; O.J. Vignoles, *Life of Charles Blacker Vignoles*, 37; May, *Charlton*, 69; and parish register for St Mary Magdalene, Canterbury, accessed via ancestry.co.uk (baptism of Charles Henry Hutton on 8 January 1800).

[2]Hutton was granted permission to retire in June 1807: Jones, *Records of the RMA*, 57.

Letter 106.
CHARLES HUTTON to JOHN BRUCE, 9 August 1813

Printed in Bruce, *Memoir*, 26–7.

London, August 9th, 1813.

DEAR SIR,

I have to thank you very much for your kind attention in sending me, from time to time, the reports of the Literary and Philosophical Society of my native town, and the sketches of Mr. Turner's philosophical lecture, all of which I find to be very interesting.[1] I have now to thank you for the 19th and 20th year's report, as well as for the 1st and 2d reports of the Jubilee School,[2] with your kind accompanying letter.

It is with truth and much satisfaction, that I have to announce my high respect, and even admiration, of both these institutions; and have to assure you, that I take shame to myself for never having contributed my mite towards the promotion of these excellent institutions.—As, however, it is better late than never to do what we think is right, I beg the favour of you to announce, that I wish to have the honour of becoming an annual subscriber of five pounds to each of them, having inclosed ten pounds to pay the first year's subscription; and will further thank you to remind me, from year to year, when each year's payment becomes due, that I may have the pleasure of sending it to you.

With much respect, I am, dear Sir,
Your's, very truly,
CHAS. HUTTON.

To Mr. John Bruce.

Notes

For John Bruce see note to Letter 104.

[1]For the Newcastle Literary and Philosophical Society see note to Letter 70; for William Turner see note to Letter 104.

[2]The Newcastle Royal Jubilee School, founded in 1810, was another project led by William Turner and John Bruce; its first and second annual reports were dated 1811 and 1812.

Letter 107.
JUDITH MILBANKE to CHARLES HUTTON, 30 October 1813

Printed in Hutton, *Recreations* (1814), vol. 4, p. 231.
Sender identified from Letter 96; also identified in 'The divining rod', *Newcastle Magazine* 2 (January 1823), 48–50 and Hutton, *Recreations* (1840), 712.

Oct. 30, 1813.

Dear Sir,

Having been from home, I did not receive the favour of your letter till yesterday, therefore take the earliest opportunity of replying to it. As I have ever had a horror of appearing in print, I had rather not have *my name* given with the account you propose of the discovering of springs; therefore, as it will be anonymous, you will judge for yourself how far you may wish to relate the circumstances with which I have acquainted you.[1] They are known to *so many*, that I am of opinion they will obtain credit in a great degree, without a *name* being formally attached to them. I can only repeat, that every experiment I have made confirms the reality of the power, though I cannot account for it.

Believe me, dear Sir, your faithful servant,

—— ——.

Note

For Judith Milbanke see note to Letter 94; see also Letters 96, 97, and 99. During the preparation of the second (1814) edition of the *Recreations* Hutton had enquired whether it would be permitted to give Lady Milbanke's name in connection with her letters about the divining rod.

[1]Hutton in fact printed the whole of Letters 94, 97, and 99, and of this letter, suppressing Lady Milbanke's name throughout.

Letter 108.
PATRICK KELLY and CHARLES HUTTON to the Committee of the Literary Fund, April 1814

London, British Library, Loan 96 RLF 1/315/2: ALS.
Only the signature is in Hutton's hand.

To the Committee of the Literary Fund
The humble Petition of Ann Coppard.
Sheweth

That your Petitioner is a daughter, of the late Mr John Robertson,[1] Author of the Elements of Navigation, & various other Scientific Works; formerly, Mathematical Master, of Christ's Hospital, afterwards Head Master, of the Royal Academy, at Portsmouth & during the last ten years of his life, Librarian to the Royal Society. That she was left a Widow, several years since, with three children, wholly unprovided for; that, by great exertion, she brought them up, & has always lived in credit, but is now, by long and severe illness, in her family, together with ⟨other⟩ unforseen misfortunes, reduced to great distress; she therefore, most humbly solicits your kind & benevolent aid, in mitigation of her sufferings, and for which, her most grateful thanks, will be due.

[*Another hand takes over:*] We the undersigned believe the above statement to be true and consider Robertson's Navigation as the most full & complete Work on that Science ever published in this Country.

Charles Hutton
P Kelly —

[*Endorsement*: April 1814 / £20 granted]

Note

Patrick Kelly (1755/6–1842), mathematician and astronomer, was also a friend of Maskelyne and Herschel (ODNB, MPHE); but Catherine Hutton, for one, disliked him: see her endorsement to Letter 127. The Literary Fund (now the Royal Literary Fund), was founded in 1790 to assist authors in need and their families.

[1]John Robertson (1707–76), mathematician: ODNB. His widow and certain of his children received support from the Royal Society after his death, but the implication of this petition is that Ann did not benefit, or was no longer able to do so by this date.

Letter 109.
CHARLES HUTTON to JOHN BRUCE, 1 July 1814

Printed in Bruce, *Memoir*, 27–9.

Bedford Row, July 1*st*, 1814.

DEAR SIR,

I am just favoured with your letter (dated as far back as April last) and the several valuable enclosures, for which you have my most sincere thanks. The three large drawings for a new method of teaching are beautifully executed, and entitled to careful preservation.[1] You have also well succeeded in their application to the rules of arithmetic, even to fractions inclusively.—The device must be a great help to very young beginners, and may, perhaps not inaptly, be called the Child's Guide to Arithmetic.

I rejoice to see so good a report of the Jubilee School,[2] and such a long and interesting list of subscribers to it. How happy the idea of thus manifesting gratitude instead of the boisterous and inordinate revelry attending illuminations! How sweet the fruits and cheering the lasting prospects of the happy substitution!

I greatly admire the three printed sketches and reports you have sent me, all very ably and judiciously written by your excellent lecturer, to whom, I am sure, the town ought in justice, to be most grateful. As you say that Mr. T.[3] is going through the last of his series of lectures, I should think it will be advisable to begin the series again, and so repeat them, with occasional variations and additions, to the edification of the successive varying audiences.—Inclosed is my annual subscription, to be equally shared as before, between the lectures and the school,[4] and continue yourself to recommend children to the school in my stead.

I must just inform you that I am now drawing fast towards a conclusion of my long protracted labours. I have got published, in three volumes of Tracts, the collection of my own original inventions and discoveries.[5]—I have also just got published, a new edition of the Recreations, in 4 vols. with some amusing additions,[6] among which is the entertaining and extraordinary correspondence with Lady Milbanke, on the Divining Rod.[7] I have also just got to press the long promised and expected new edition of my Dictionary,[8] with great improvements, which have cost me immense labour. It will still be in 2 vols. a little larger than before, and to be published altogether. After that, I think there will only remain to incorporate, in regular order, the subjects of my Course of Mathematics, all the three volumes into two large ones, on the occasion of some future edition.[9]

With much regard, I am,

Your's, very truly,

CHAS. HUTTON.

To Mr. John Bruce.

Notes

For John Bruce see note to Letter 104.

[1]According to the surrounding text in Bruce, *Memoir*, this referred to an essay on Pestalozzi's system of education, which Bruce had presented to the Newcastle Literary and Philosophical Society; it seems to have remained unpublished.

[2]For the Newcastle Jubilee School see note to Letter 106.

[3]William Turner, the 'excellent lecturer' of the previous sentence, for whom see note to Letter 104.

[4]See Letter 106.

[5]Hutton, *Tracts* (1812).

[6]Hutton, *Recreations* (1814).

[7]That is, Letters 94, 97, 99, and 107. For Judith Milbanke see note to Letter 94.

[8]Hutton, *Dictionary* (1815).

[9]The first edition of Hutton's *Course* appeared in 1798; a third volume jointly authored by Hutton and Olinthus Gregory was added in the sixth edition of 1810–11. No reorganized two-volume edition of the kind described here appeared during Hutton's lifetime.

Letter 110.
JOHN VIGNOLES to CHARLES HUTTON, 21 October 1814

Portsmouth, History Centre, Vignoles Papers, 1072A/75/1: ALS.

Letter 110. John Vignoles to Charles Hutton, 21 October 1814

Holt 21st Octr 1814

Sir

This morning I was favored with yours of yesterday's date and hasten to ⟨give⟩ such information as is within my power. Altho' no way related I well know who the Captn. Vignoles you mention[1] was. My[a] Uncle and Father were Majors to the two Batt^{ns}. of the 31st. Reg^{t}. the Elder, my Uncle, to the first, my Father to the second; at which time M^{r}. Francis Vignoles was a Lieut^{t} & a Captn. in that Reg^{t}. and went to America with that Regiment where there were lands alloted to him and which, I believe, remain still unclaimed; I had this latter intelligence from Major Pilot[2] who was with the Reg^{t}. at (I think) at Pensacola; but Major Pilot who has been for some time a resident in Bath, can fully inform you respecting; however I must add that ⟨he⟩ has not deem'd them worth claiming; yet perhaps a claim of that kind might ⟨be⟩ worth something to a young man who has his fortune to make — In respect to the Isle of Man I can set you right; a claim there was set up by me; a Mr Bonvillete who was established there as Marchan[3] died intestate; he had had a sister married to my Father, and she having some property at her disposal and ⟨made me⟩ her universal Heir, but died before her Brother, when he died no claimant appeared until I sent over one who I thought had a sufficient claim but was not admitted, a friend of mine then desired me to hand my claim to him and he hoped he would get it for me, I did so but it was too late time had elapsed, and it had Escheated to the Family. I[t] was by that means the name of Vignoles was mentioned. As to French Claims, forty years ⟨ago⟩ when I was in France I found that all such claims were nugatory. Perhaps you may recollect a young man of that Name of Vignoles admitted at Wollwich shortly before you Sir left the Academy ⟨(<u>He was my Son</u>)⟩[b],[4] I then wrote to some Friend of mine to communicate to you the Pensacola business but I really do not recollect who. To have given you more favorable intelligence would have been highly pleasing to Sir,

Your Obedient Servant
J^{no} Vignoles

D^{r}. Hutton
Bedford Row
London

---

[a] *altered from*: His.

[b] There is no indication of where this insertion should be placed.

Notes

John Vignoles is not known except for this and Letter 111; evidently Hutton had approached him in the belief that he was a member of, or otherwise had knowledge of, the family of Charles Henry Vignoles (for whom see note to Letter 49).

[1]Hutton had inquired about Captain Francis Vignoles, the father of Charles Henry Vignoles: see K.H. Vignoles, *Romantic Engineer*, 22.

[2]I have not been able to identify a 'Major Pilot'. Perhaps the most likely candidate is the (by this date) Lieutenant Colonel Henry Powlett, who served in the West Indies with the 60th Foot from 1789 to 1792. He was placed on half pay after the 7th Garrison Battalion was disbanded in 1810 (subsequently, from 1815, he served with the 5th Royal Veteran Battalion); I have not been able to confirm that he resided in Bath during this period: Philippart, *Royal Military Calendar* (1815–16), vol. 3, p. 365.

[3]'Marchan' is a Manx word meaning merchant.

[4]The only Vignoles admitted to the Royal Military Academy about this time was Samuel Vignoles in 1811; perhaps John Vignoles was mistaken about the date (1807) at which Hutton left the Academy. See also Letter 111, where it was mentioned that in 1813 a commission in the 41st Regiment of Foot was announced for John Vignoles' son, but that he was not permitted to leave Woolwich to take it up. He was in the event commissioned in the Royal Artillery in 1815: information from the Sandhurst registers via archive.sandhurstcollection.co.uk/search.

Letter 111.
JOHN AND ANNA VIGNOLES to CHARLES HUTTON, 20 January 1815

Portsmouth, History Centre, Vignoles Papers, 1072A/79: ALS.

Overbury — Tewkesbury
20th. Janry [*in another hand: 1815.*]

Sir,

My object in addressing you is to express my satisfaction that an act of kindness intended[a] for my Son,[1] but which the military line he has embraced rendered him incapable of profiting by should have turned so much to the advantage of your Grandson;[2] and more particularly, as he appears from unequivocal testimony to be so deserving it. Your[b] feelings I can appreciate,

[a] *deleted*: intended.
[b] *deleted*: feeli.

on seeing the ofspring of your Child not only prosper but applauded. writing is too painfull to me, to admit of my detailing the circumstances of this transaction, and therefore shall hand my pen over to M^{rs}. Vignoles who will explain it to you,

I am S^{r}. your very obed^{t}. s^{t}

Jno Vignoles

[*The hand changes to that of—presumably—Mrs Vignoles.*]

Sir

if <u>your</u> Grandson is in the Royals & in America I hope I have been the Means (without any merit of my own, tho' well disposed to give assistance to any one who wants it) & shall Mention to you a paragraph I had lately in a letter from a Friend who imagined I had a Son.in the Army & in[a] America & wishing to serve me had recommended him to Coll. Muller[3] previous to his Embarkation for America – & since his arrival there she has had a letter from Coll. Muller from which she transcribes the following passage

"shou'd you meet Mr Vignoles's Mother tell her that her Son has in that Unfortunate Ship wreck given me fresh proof of zeal & devotedness to the service, ⟨& his ?particular duty⟩, & for which I have not failed to recommend him for a Staff appointment to the Commander in Chief (where Coll. M is)["][4]

the Son whom my Friend supposed was in America – had (two Years since) been gazetted for a commission in the 41st. at Canada – but as he had been placed at Woo[lwich] the Governors were highly disple[ased?] at thinking of his removal & ?woul[d] ?not allow him to accept of it[b] – & <u>there</u> he still is without any chance of promotion.

However I may lament that <u>my</u> Son could not Avail himself of the kindness of Friends give me leave to congratulate you on what I hope may turn out to the advantage of your Grandson & am with good wishes Much yours

?Anna Vignoles

Doctor Hutton
34. Bedford Row
Bedford Lane
London

---

[a] *altered from:* at.

[b] A combination of unclear handwriting and loss of text due to the seal makes this passage difficult to decipher.

Notes

For John Vignoles see note to Letter 110.

[1]See note to Letter 110.

[2]For Charles Blacker Vignoles see note to Letter 61. By this date he was pursuing a military career and had been commissioned in the first regiment of foot. After action in the Netherlands, parole, and exchange he travelled—as described later in this letter—to America. His clandestine engagement to Mary Griffiths, a penniless ward, was not yet known to his family. K.H. Vignoles, *Romantic Engineer*, 9–12.

[3]Lieutenant Colonel Frederick Muller of the first (Royal Scots) Regiment of Foot: War-Office, *Army List*, 29, 138; Philippart, *Royal Military Calendar* (1820), vol. 4, p. 361.

[4]Charles Blacker Vignoles was indeed involved in the wreck of the *Leopard* at the mouth of the St Lawrence river in June 1814; this appears to be the only report that he distinguished himself on that occasion. No staff appointment in fact followed, despite Vignoles' repeatedly stated wish to serve as a staff engineer, until he became ADC to General Sir Thomas Brisbane at Valenciennes in April 1816: K.H. Vignoles, *Romantic Engineer*, 12.

Letter 112.
CHARLES HUTTON to JOHN BRUCE, 1816

Extracts printed in Bruce, *Memoir*, 30–32.

As you say, my good friend, it is a very alarming circumstance, that in your Schoolmasters' Association,[1] consisting of only forty-nine members, you have nineteen widows to support or to assist; and, perhaps, not fully to be accounted for. Probably it may be partly accounted for, from the circumstance that members, after becoming widowers, often marry again, and commonly with women (often much) younger than themselves; after which, it is usual for the husband to die before the second wife.[2] Hence it must, and does, happen, that there is always existing in society more widows than widowers; it being much easier for a man to get a second wife, than for a woman to get a second husband. I believe this will account for the situation of your Association. The same thing in some degree happens in all such societies, but then they commonly balance that cause by imposing a proportionate fine on all repeated marriages of their members. The circumstance, then, of your widows ought not to be alarming to schoolmasters, as if their occupation is an unhealthy one, or that they are not usually long-lived; for the fact is, that they are more long-lived than most other classes, which happens partly on account of the nature of their professions, but still more from their moderate and moral habits, being the most regular and virtuous class of men in society, as might easily be proved by many cogent reasons. Inclosed is the sum of £5. by way of beginning my annual subscription for your Schoolmasters' Association, and possibly I may hereafter do something more for it also.[3]

...

Through the favour of Mr. Jack,[4] I last evening received your interesting packet, and the several satisfactory documents of useful information, relating to the many interesting literary concerns in which my natal place is so laudably employed, and which transmissions, I trust, you will have the goodness to continue during the small expected remainder of my life.

I observe the report of the Literary and Philosophical Society,[5] and am rather surprised the Society has not kept and appropriated all the papers that have been read at the meetings, which I believe, is the uniform practice of all other societies, and out of which communications a selection is made for publication, at certain periods, or when convenient.

Your schools I am very well pleased with, as the enlightening the lower orders, as well as the higher, is the way to render them virtuous, happy, and respectable. I sent a few days ago my annual subscription to Mr. Atkinson,[6] £5. to the school, and the same sum to Mr. Turner's lectures,[7] which I hope, he will contrive to make perpetual, as there is no end to their usefulness.

Notes

For John Bruce see note to Letter 104.

[1]Formed in 1774, the Newcastle Schoolmasters' Association provided support for 'aged and infirm members and widows': Bruce, *Memoir*, 29. Bruce had mentioned to Hutton that 'owing to the great number of widows who are claimants upon the funds of the institution' it was unable to make payments of the size the founders had intended, and asked him (p. 30)

> whether the great proportion of widows at that time receiving the stated benefit from the Association, (nineteen widows to forty-nine members) was not a proof that the occupation of a schoolmaster was less healthy than some other professions?

This section of his letter was his reply.

[2]Hutton himself had been widowed and remarried to a woman somewhat younger than himself: for Margaret Hutton (née Ord) see note to Letter 31. This passage may suggest that at this stage he expected Margaret to outlive him.

[3]Hutton enclosed a further sum of £5 for the Schoolmasters' Association with Letter 116, and subsequently, as detailed in that letter, made his subscription perpetual.

[4]Possibly the proprietor of Jack's long-lived coffee-house in Soho; compare Letters 28 and 29.

[5]For the Newcastle Literary and Philosophical Society see note to Letter 70.

[6]'Atkinson' was most probably the Newcastle-based mathematician and teacher Henry Atkinson (1781–1829: Welford, *Men of Mark*, vol. 1, pp. 140–3; ODNB), and the implication would be that he was at this time associated with the Newcastle Royal Jubilee School, to which Hutton was a subscriber (see Letter 106 and notes); I have not seen other evidence for this. Hutton's reference to 'schools' also included the successful Percy Street Academy of which John Bruce was the founder and director; while the Jubilee School educated the lower orders, Bruce aimed to offer the higher an education comparable to that of the ancient public schools: Welford, *Men of Mark*, vol. 1, pp. 410–11.

[7]For William Turner see note to Letter 104.

Letter 113.
CHARLES HUTTON to JOHN BRUCE, 15 January 1816

Extract printed in Bruce, *Memoir*, 29.
Date from surrounding text.

[...] it would certainly, my dear Sir, give us[1] much pleasure to visit our old town, and all the well-known scenes in the environs, and our few good friends; but it seems a great undertaking for persons who move so little, and who feel that every year adds to their unfitness for the undertaking. However, I do not quite despair of the so-long-desired pleasure.[2]

Notes

For John Bruce see note to Letter 104.

[1]It is not clear whether 'us' was meant to include Hutton's wife Margaret, aged about 63, or his daughter Isabella, aged 53, or indeed both: see notes to Letters 31 and 16 respectively. The implication that the infirmities of advancing age were the main obstacle to a visit may suggest Margaret, strengthening the possibility that she had lived in Newcastle.

[2]On Hutton's projected visit to Newcastle see also Letters 115, 129, and 130.

Letter 114.
CHARLES HUTTON, testimonial for CHARLES BABBAGE, 21 March 1816

London, British Library, Add. MS 37182, fols 60–61: ALS.

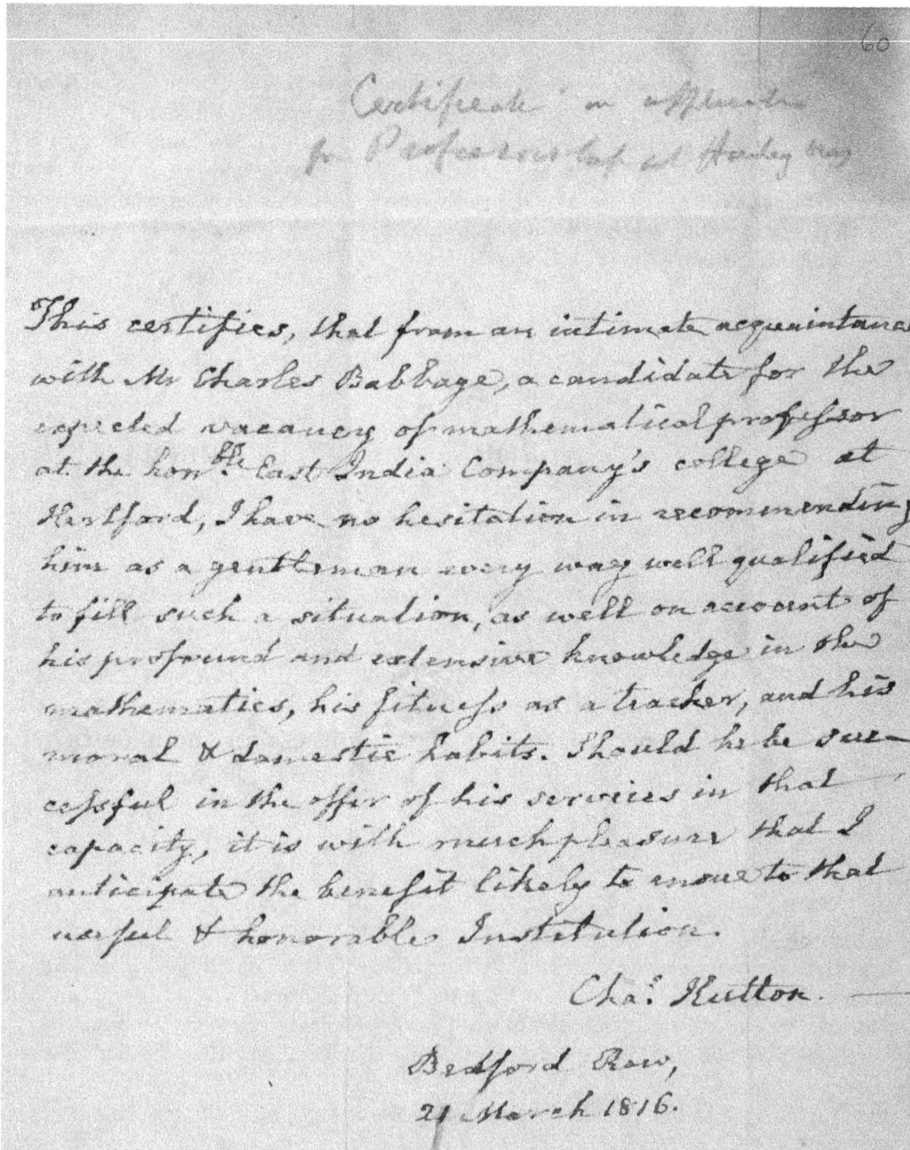

This certifies, that from an intimate acquaintance with Mr Charles Babbage, a candidate for the expected vacancy of mathematical professor at the hon^{ble} East India Company's college at Hertford, I have no hesitation in recommending him as a gentleman every way well qualified to fill such a situation, as well on account of his profound and extensive knowledge in the mathematics, his fitness as a teacher, and his moral & domestic habits. Should he be successful in the offer of his services in that capacity, it is with much pleasure that I anticipate the benefit likely to ensue to that useful & honorable Institution.

Chas. Hutton.

Bedford Row,
21 March 1816.

Figure 4: Letter 114: Charles Hutton, testimonial for Charles Babbage, 21 March 1816. British Library, Add. MS 37182, fol. 60^{r}. ©The British Library Board.

Bedford Row,
21 March 1816.

This certifies, that from an intimate acquaintance with Mr Charles Babbage, a candidate for the expected vacancy of mathematical professor at the hon[ble]. East India Company's college at Hartford,[1] I have no hesitation in recommending him as a gentleman every way well qualified to fill such a situation, as well on account of his profound and extensive knowledge in the mathematics, his fitness as a teacher, and his moral & domestic habits. Should he be successful in the offer of his services in that capacity, it is with much pleasure that I anticipate the benefit likely to ensue to that useful & honorable Institution.

Cha[s]. Hutton.

Note

Charles Babbage (1791–1871) went on to be well known and prolific as a mathematician and computer pioneer. He had left Cambridge in 1814 and lived in London from late 1815; on 14 March 1816 he had been elected a Fellow of the Royal Society and in 1820 he would participate in the foundation of the Astronomical Society, of which Hutton was also a member. He was a friend of John Herschel (see note to Letter 132) and worked with Francis Maseres (see note to Letter 7) on Maseres and Babbage, *Scriptores optici* (a remark in Dauben and Scriba, *History of Mathematics*, 163 connecting that work with Hutton seems to me to be in error): ODNB, DSB, MPHE. Nevertheless, this testimonial seems to be the only known direct evidence for Babbage's supposedly 'intimate' acquaintance with Hutton. A report in the *Caledonian Mercury* (2 September 1819) stated that Hutton also supported Babbage in an application for the mathematical chair at Edinburgh.

Babbage would later become close to Ada Lovelace, the granddaughter of Judith Milbanke (for whom see note to Letter 94), but it is not known that their mutual acquaintance with Hutton played any part in this.

[1]Babbage was not in fact appointed to the position at Hartford.

Letter 115.
CHARLES HUTTON to JOHN BRUCE, first half of 1816

Extract printed in Bruce, *Memoir*, 32–3.
Date from internal evidence.

My library, yes, I have been cruelly prevented from having it kept together in the British Museum, by my old implacable enemy, the President of the Royal Society.[1]—So it is to be sold by auction, as I shall have little or no further use for it, and it would prevent me from chusing another residence, and as it could be of no use to any of my family, after my death.[2] The dispersion is a great pity, as it is esteemed the best mathematical collection in the kingdom. By this time, I believe, Mr. Charnley has received some of the catalogues, and I have here inclosed one for yourself.[3] Possibly some articles might be picked out, that might be desirable for the library of your Society.

My good Sir, both Mrs. H. and myself have suffered very much in our health this last winter and spring, more so indeed than in several years before, which has rendered me quite the old man, a sensation I hardly felt before; so that I fear I must say it is rather improbable that I shall be able to visit Newcastle this year, though I so much wish to do it.[4]

Notes

For John Bruce see note to Letter 104. Hutton's library was sold in June 1816, and this letter evidently preceded the sale by only quite a short time.

[1]For Joseph Banks see note to Letter 22. A fuller account of the affair appeared in [Gregory], 'A Review', 248–9, where it was asserted that the trustees of the British Museum had agreed to purchase Hutton's library, and that matters had proceeded even as far as the taking of an inventory and negotiations about a purchase price when Banks—one of the trustees, but previously absent in the country—intervened to block the arrangement.

[2]The sale took place from 11 to 17 June 1816: *A Catalogue*, title page. It is not really clear why Hutton responded to the failure of the sale to the British Museum by dispersing his library at a public auction; as he would hint in Letters 116 and 129 and as Bruce had apparently indicated to him, the Newcastle Literary and Philosophical Society would have been one possible destination for the collection, and there were certainly others. Nor did Hutton, as he hinted here and on the title page of *A Catalogue*, move house after the sale.

[3]A 'Mr. Chandley' of Newcastle-upon-Tyne was named on the title page of *A Catalogue* as one of the sellers of that catalogue; the spelling 'Charnely' was used in the biography of the father and son William and Emerson of that name, booksellers, in Welford, *Men of Mark*, vol. 1, pp. 541–5. It was the son, Emerson, who was referred to here.

[4]On Hutton's projected visit to Newcastle see also Letters 113, 129, and 130. It was reported in Bruce, *Memoir*, 44 that his London friends finally dissuaded him from the plan, fearful for the effect on his health.

Letter 116.
CHARLES HUTTON to JOHN BRUCE, 8 May 1817

Printed in Bruce, *Memoir*, 33–4.

Bedford Row, May 8th, 1817.

MY DEAR SIR,

Since the date of your last letter, which is a long time ago, I have had almost continual ill health, which has reduced me to such a condition, that I can hardly write my own name. You give me much satisfaction about the books you had out of the sale of my library, and would to God that you had got the whole.[1] Inclosed is a bank note of £15., viz. £5. for Mr. Turner's Lectures, £5. for the Jubilee School, and £5. for the Schoolmasters' Association.[2] And as my health is now so precarious, I wish to fall upon a method of making the two latter perpetual, and leave it to you to contrive the method and to transact it. My idea is to sink £100. at common and perpetual interest, for each of these two concerns. Perhaps those £200. could be lent in that way to the town Corporation, on such permanent and safe manner.[3] As to the lectures, I shall continue to pay that annually.

But the grand misfortune of all that has befallen me, is the recent loss of my dear and valuable wife,[4] which has left me almost alone in the world.

With much esteem and regard,
I am, dear Sir,
Your's, very affectionately,
CHAS. HUTTON.

To Mr. Bruce.

Notes

For John Bruce see note to Letter 104.

[1]See Letter 115.

[2]See Letters 106 and 112.

[3]It was reported in Bruce, *Memoir*, 34–6 that Hutton did indeed make an arrangement of this kind through the town clerk, the Corporation of Newcastle accepting £200 and agreeing to pay perpetually £5 annually to each of the two institutions.

[4]For Margaret Hutton see note to Letter 31; she died on 15 March 1817.

Letter 117.
CHARLES HUTTON to JOHN BRUCE, 24 April 1818

Extract printed in Bruce, *Memoir*, 38.

I am going to try to scrawl for you two or three lines, though still so weak, that my fingers are not able to command the pen.[1] I have been for eight or nine months closely confined in the house, good part of the time in bed, and all to the chamber, a good while without hope, owing to the severity of cold upon cold. I am now, however, very slowly recovering, and seem only to want fine and warm weather, to enable me to get out a little, and to enable me to go on a little while longer. I thank you, dear Sir, very much for your kind remembrances of me, and for the pleasing pamphlets on the Literary Society, and the Jubilee School,[2] the improving state of which I greatly rejoice at, as well as the improving state of comfort in the town, by the use of gas lights. In most of my time, the town was in total darkness, and only received the street oil lamps, a very short time before my departure from it.

Notes

For John Bruce see note to Letter 104.

[1]Bruce, *Memoir*, 37 reported that in another letter of presumably around the same time Hutton said it was 'now no small task to scratch a few lines of a letter'.

[2]For the Newcastle Literary and Philosophical Society see note to Letter 70; for the Jubilee School see note to Letter 106.

Letter 118.
CHARLES HUTTON to JOHN BRUCE, ?May 1818

Extract printed in Bruce, *John Collingwood Bruce*, 13. Date from surrounding text. It is possible that this is another passage from Letter 117.

I cannot recommend a school for your son; they are all so bad that I would not venture to do it. I conversed with our friend Dr Gregory[1] about it, and he seemed to have one in view which he intended to mention to you.

Note

For John Bruce see note to Letter 104; he had sought Hutton's advice about a school for his son, John Collingwood Bruce (1805–92). The latter became acquainted with Hutton after his move to London in 1819: ODNB; Bruce, *John Collingwood Bruce*.

[1]Olinthus Gilbert Gregory (1774–1841), mathematician, was Hutton's successor both as editor of the *Ladies' Diary* and other almanacs and, eventually, as Professor of Mathematics at the Royal Military Academy; he was also the recipient of some of Hutton's books and manuscripts during the final period of his life and one of his most informative obituarists: ODNB, and see the introduction to this volume as well as Gregory, 'Memoir' on his relationship with Hutton. Gregory's advice reached Bruce in a letter dated 12 May 1818; he recommended the Protestant Dissenters' Grammar School at Mill Hill in Middlesex, which J.C. Bruce went on to attend: Bruce, *John Collingwood Bruce*, 14.

Letter 119. CHARLES HUTTON to CATHERINE HUTTON, 26 January 1819

London, Wellcome Collection, MS 5270 no. 35: ALS.

Bedford Row
26 Jan^{ry}. 1819.

My Dear Miss Hutton,

I have been favoured, and very agreeably surprised, with the first volume of a series of travels, all over the world I hope, in the same manner as the present beginning.[1] It is needless to say how much I admire it, after having devoured every word of it, from beginning to end. The plan is admirable, and original, at the same time so simple and obvious, that one wonders it has never before been adopted, after so many brains have been racked for new inventions. The execution too is not less excellent: the language natural, easy, correct, and well appropriated to the subject; concise, yet sufficiently copious; so that a reader of the narrations could never doubt their originality, had he not been acquainted with the originals, and had you not proclaimed the fiction. The address too is admirable in glozing over certain scenes of galantry with which ⟨Bruce⟩[2] had been indulged. The map too is very useful, and neatly executed, except that the seas are too darkly shaded. — Perhaps there is a press error in the latitude of Suakem (3° 14′) near the bottom of pa. 226.[3] — In the further progress of your travels, when about to commence a particular rout, it may be well to inform your bookseller what authors you possess on that rout, and request him to procure for you any others that he may know of. I do not recollect the names of any travellers beyond Abyssinia southward along the east side of Africa to the Cape. About the Cape there are many, and thence northward along the west side. I have read a good account of the kingdom of Dahouny on the gold coast, but have forgot the author's name. You have Alli Bey for Morocco &c, and still better a Mr Lempriere, a surgeon who was invited from Gibraltar into Morocco by the emperor, to cure the prince his son, which you ought to have.[4] — And be sure never omit to give maps of the rout you travel.

But what shall I say, dear Coz, to the excellent and flattering letter with which the book is accompanied. It is but a poor expression of my feelings to say it has given me great pleasure; in short I am quite transported, and shall carefully preserve it as a curious and sacred deposit to be transmitted to my latest posterity. But you have artfully, though fairly, taken [the] most effectual way to procure a specimen of my [pre]sent hand writing, if writing one may call [su]ch a tottering scrawl, and a thing which I have not attempted for a long time past, as your dear sister here is my constant amanuensis on all occasions.[5]

It is a 1000 pities your book is not faced with a good portrait of the author, whose features will make an excellent one: you cannot be excused: it must be given in the 2d vol. otherwise you will never be forgiven, as the world demands it. But as these things are not so well done in the country, you must come up here when the season is favourable, and stay a few days with your cousins, when we will have the artists to come and take both yours & mine, even in my hair as you say, such as it is.[6] To this you must make no objections, it being of all things most proper to be done.

My son, with his wife and boy, are very well, and passing their winter at Aberdeen.[7] My dear companion here[8] is also quite well, and unites in kind regards to her dear sister, with,

Dear Madam,

your affectionate hble Servt.

Chas. Hutton.

PS. Pray forgive the blunder I have made in deranging the proper succession of the pages, as it is the first fault of the kind I ever committed.[9]

Miss Hutton
Bennett's Hill
near Birmingham

Notes

Catherine Hutton (1756–1846) was prominent during her lifetime as a novelist and letter-writer: ODNB, Hutton, *Reminiscences*; see also Letters 121, 126, 128, and 131. It is not known when Catherine and Charles Hutton became acquainted; the 'dear Coz' of this letter, together with the reference to Isabella Hutton as 'your dear sister', seems to suggest an acquaintance of some standing, and her biographer mentioned visits to his home at which she met 'learned men' including Maskelyne (who died in 1811): Hutton, *Reminiscences*, 180. Catherine was an inveterate autograph collector and celebrity hunter and may well have instigated the correspondence after Hutton had become a prominent author.

Charles and Catherine regularly used the word 'cousin' in their correspondence, but there is no evidence that the two believed there was really a family relationship; a later newspaper report about Catherine and Hutton's daughter Isabella (who continued to correspond) stated that the two ersatz sisters were 'unconscious of actual consanguinity': *Monthly Magazine* 26 (August 1838), 206–7; compare Jewitt, 'Pedigree of the family of Hutton, of Derby, Birmingham', which shows no link with Charles.

[1]Hutton referred to Hutton, *Africa*, a fictional travel narrative.

[2]That is, the traveller and travel writer James Bruce of Kinnaird (1730–94: ODNB); his *Travels* included, as Nigel Leask remarks in the ODNB, 'Shandean asides boasting of his success with African women'.

[3]Catherine Hutton referred at the place indicated to what seems to have been the town of Sawākin; the latitude intended was perhaps 37°14′.

[4]'Bey', *Travels*; Lemprière, *Tour*.

[5]A number of Hutton's late letters are indeed in what appears to be Isabella's hand (Letters 102, 121, 123, and 132); but compare the implication in Letter 117 that Hutton was there writing with his own hand.

[6]An engraving of Catherine Hutton by William Read was published with *La Belle Assemblée* 29 (June 1824), and an engraving of Charles Hutton (in his hair) by the same artist is extant (Wellcome Library, Iconographic Collections, ICV No. 3190). It is therefore possible, though not certain, that co-ordinated sittings did in fact take place. See also Letter 121.

[7]For Henry Hutton see note to Letter 12; the reference was to his second wife, Miss Barlow, and his second son, Henry (1809–63): ODNB *s.v.* Charles Hutton; also information from the censuses of 1851 and 1861.

[8]For Isabella Hutton see note to Letter 16.

[9]Hutton wrote successively on fols 1[r], 2[v], and 1[v], and then folded the sheet back on itself to make 2[r] the outside wrapper, which he addressed and sealed.

Letter 120. CHARLES HUTTON to PIERRE-SIMON DE LAPLACE, 9 April 1819

Printed in the *Journal de physique, de chimie, et de l'histoire naturelle* 90 (April 1820), 307–12.
English text printed in the *Philosophical Magazine* ser. 1, 55 (February 1820), 81–5, and reprinted verbatim as a separate pamphlet.

Monsieur le Marquis,

J'ai désiré pendant long-temps et avec beaucoup de force, d'avoir l'honneur de vous offrir mes respects en personne; mais comme il n'est guère probable que je puisse désormais faire un voyage en France, étant maintenant dans ma quatre-vingt-deuxième année, il ne me reste qu'à m'adresser à vous par écrit, et qu'à vous prier d'accepter l'hommage d'un sincère partisan de ces sciences à l'avancement desquelles vous avez contribué si puissamment avec le plus grand honneur pour vous-même, aussi bien que pour l'avantage du monde savant.

Je profite d'une occasion qui m'est offerte de faire parvenir cette lettre à Paris, pour vous prier en même temps d'accepter un exemplaire de mes traités sur différens sujets physiques et mathématiques,[1] que je demande de vous présenter comme un foible, mais sincère gage du profond respect qu'avec le reste du monde, j'ai pour votre génie et vos talens.

Parmi ces traités, je désire particulièrement d'attirer votre attention sur celui qui commence le second volume, sur la densité de la terre, et de vous exprimer mon désir que, vous qui avez déjà fait des recherches si profondes sur sa figure et sur la doctrine des marées, vous vous occupiez de sa densité, afin de corriger mes erreurs, ou de confirmer la vérité de mes travaux. En vérité je suis surpris que de semblables expériences et recherches n'aient pas encore été faites jusqu'ici en France, où la science, dans toutes les autres branches, a été poussée avec le zèle le plus louable et les succès les plus remarquables.

Comme vous devez naturellement prendre intérêt à tout ce qui a rapport à une entreprise aussi importante, qu'il me soit permis de rapporter ici certaines circonstances qui y ont rapport, avec une idée générale de la manière dont les opérations furent conduites, alors je terminerai, en relevant une légère erreur de la Connaissance des Tems pour 1821, pag. 330, ligne deux,[2] où mes travaux, pour la solution de cet important problème, sont attribués au Dr Maskeline;[3] chose qu'il n'a jamais réclamée lui-même, et qu'au contraire, il a toujours relevée comme une erreur, toutes les fois qu'il l'a entendu commettre devant lui.

Un peu avant l'année 1769, lorsqu'on attendoit un transit de la planète Vénus, la Société royale sentit l'utilité qu'il y auroit d'envoyer plusieurs habiles astronomes dans des lieux différens, pour l'observer; mais comme la dépense de semblables missions étoit au-dessus de ses moyens, elle fit connoître la difficulté au roi, et S. M. voulut bien accorder une somme considérable pour en faire les frais. En conséquence, deux astronomes de mes amis, MM. Wales et Bayly,[4] furent en temps convenable envoyés à la baie d'Hudson, où ils passèrent un été et un hiver très-rigoureux; à leur retour en Angleterre, ils donnèrent l'histoire de leur mission. Elle fut communiquée au gouvernement, avec un mémoire des dépenses faites; et comme elles se trouvèrent inférieures à la somme qui avoit été allouée, on demanda au roi ce qu'il lui plaisoit qu'il fût fait du reste, et il répondit que la Société royale l'emploieroit pour telle expérience ou tel sujet qui lui paroîtroit convenable.

Après y avoir suffisamment réfléchi, la Société royale prit la résolution d'employer cette somme à déterminer, par expérience, l'attraction réciproque et universelle de la matière, en observant la déviation de la ligne d'à-plomb de la perpendiculaire par l'attraction de quelques montagnes: résolution à laquelle le Dr Maskeline prit la principale part. En conséquence, une correspondance fut établie avec plusieurs personnes dans toutes les parties de la Grande-Bretagne, pour découvrir une montagne convenable à l'expérience. On s'arrêta au mont Schichallien, dans le centre de l'Ecosse.[5] Le second point étoit de savoir quelle personne on prendroit pour faire l'expérience: on désiroit beaucoup que le Dr Maskeline voulût bien s'en charger; mais il s'en excusa, et donna pour raison que son devoir d'astronome royal ne lui permettoit pas de s'absenter. Après quelques délais, on s'accorda pour envoyer M. Burrow, qui avoit été dernièrement aide de M. Maskeline à l'observatoire royal.[6] Ses instructions portoient, premièrement, de faire un examen soigné de la montagne dans toutes ses dimensions, horizontalement dans toute direction, et de faire un grand nombre de sections verticales de tous côtés, de manière à ce qu'on pût en avoir un parfait modèle, et que sa grandeur pût être estimée avec exactitude; après quoi il devoit observer la déviation de la ligne à-plomb. Après un an ou deux employés à ces mesures préliminaires, la Société craignant que l'ouvrage ne fût pas bien exécuté, sollicita M. le Dr Maskeline d'y aller lui-même, et de diriger les opérations; et, en effet, après avoir obtenu la permission du roi, il y fut, et rapporta les mesures et les descriptions de M. Burrow, avec ses propres observations sur la déviation du fil à-plomb, faites des deux côtés de la montagne. Ces opérations employèrent les années 1774, 1775 et 1776; et la méthode d'exécution qu'il avoit suivie fut exposée par le Dr Maskeline, dans les Transactions philosophiques, pour l'année 1775.[7]

Ainsi fut résolu autant qu'il pouvoit l'être expérimentalement le problème, qu'*il y a une attraction mutuelle dans toute la matière,* et que la somme des deux attractions de chaque côté de la montagne en particulier, montoit à 11,6 secondes. La Société fut extrêmement satisfaite de ces opérations. On s'aperçut aussi quel usage ultérieur on pouvoit faire de ces mesures; et l'on vit que par les mêmes moyens, après deux calculs faits, la densité moyenne de la terre pouvoit être déterminée: car on avoit maintenant ces données, c'est-à-dire la masse et l'attraction d'une montagne, avec la grandeur et l'attraction de la terre, pour déterminer sa densité et sa masse. On étoit donc sur la voie d'un grand et fort important objet; mais la question étoit de savoir *qui* feroit les calculs immenses dont on avoit besoin; qui le *voudroit*, et même qui le *pourroit*. Il pouvoit y avoir une ou deux personnes que l'on pensoit capables d'entreprendre cette tâche; mais elles, et tous les autres membres de la Société, restoient effrayés à la vue du nombre et de la grandeur des calculs, et reculoient devant l'immensité du travail; ce fut alors qu'on jeta les yeux sur moi, et que le président et le conseil de la Société me demandèrent d'entreprendre cette tâche, ce à quoi je fus aussi fortement sollicité par le Dr Maskeline. Etant alors un *jeune membre de la Société,* et désireux de me distinguer, j'entrepris l'opération, et devins responsable du résultat, qu'après un travail journalier de près de deux ans, je présentai à la Société, avec un détail [m]inutieux des particularités des calculs, accompagné de dessins convenables. La Société fut si satisfaite, qu'elle me décerna un prix, et imprima mon Mémoire dans les Transactions philosophiques pour 1778, exactement dans la forme où je le lui avais remis.[8]

Il faut observer que, dans cette opération, la première de cette espèce qui eût encore été faite, tous les procédés de calculs étoient à imaginer ou à inventer, et que sans plusieurs plans que je pratiquai, il est fort douteux que le travail eût pu jamais être exécuté avec soin. J'apporterai, comme preuve de la vérité de ces assertions, le témoignage le plus honorable qui en a été donné par l'un des meilleurs mathématiciens du royaume, le professeur Playfair.[9] Le résultat de mes calculs fut que la densité moyenne de la terre est à celle de la montagne, comme 9 : 5. La densité de la montagne n'étoit pas connue alors, et par conséquent celle de la terre comparée à celle de l'eau, ne pouvoit alors être déterminée. Mais pour montrer la manière dont la comparaison pouvoit être faite, je pris pour exemple la plus petite densité de la pierre $2\frac{1}{2}$, quoique ce fût évidemment au-dessous de la densité réelle de la roche de la montagne; alors $\frac{9}{5} \times \frac{5}{2} = 4\frac{1}{2}$, devoit être moindre que la densité de la terre, comme le nombre adopté, jusqu'à ce que nous puissions connoître la densité réelle de la montagne.

Plusieurs années après, le professeur Playfair fit un examen lithologique de cette montagne, et il trouva qu'elle étoit formée de roches de densités très-différentes, depuis 2,5 jusqu'à 3,2, dont la moyenne est d'environ $2\frac{3}{4}$, ou 2,8; prenant le premier nombre $2\frac{3}{4}$ comme la densité de la montagne, alors $\frac{9}{5} \times 2\frac{3}{4} = \frac{99}{20} = 4\frac{19}{20}$[a] ou presque 5, est la densité moyenne de la terre. C'est ce résultat que j'ai substitué au premier, à sa place, dans l'abrégé que j'ai donné des Transactions philosophiques, depuis son origine, jusqu'à 1800, en 18 grands volumes in-4°, comme on pourra le voir.[10] Si nous prenons 2,8, au lieu de $\frac{3}{4}$ employé ci-dessus, alors $\frac{9}{5} \times 2,8 = 5,04$, est la densité de la terre, un peu au-dessus de 5, comme le premier résultat étoit un peu au-dessous. C'est pourquoi je me déterminai à regarder le nombre 5 comme le résultat de la densité moyenne de la terre, déterminée par mes calculs. J'ai également donné la même conclusion dans le premier mémoire du second tome du recueil de mes traités, où tout le procédé est exposé au long,[11] comme on pourra le voir dans le choix de ces traités, qui vous a été remis avec cette lettre, et que j'ai aussi présenté dernièrement à l'Institut royal.[12]

---

[a]Source has + in place of ×.

Maintenant, après des preuves évidentes que j'ai été le premier et le seul, qui ait calculé la moyenne densité de la terre d'après les mesures originales, et cela depuis près d'un demi-siècle, on doit être étonné de voir que l'honneur de cette détermination est sur le point d'être transféré à une autre personne qui ne l'a jamais désiré et n'y a jamais pensé, à moins que l'erreur accidentelle, ou la méprise introduite dans la *Connaissance des Temps*, ne soit reconnue et corrigée. C'est pourquoi j'espère avec confiance de la libéralité et de l'honneur bien connus de votre caractère, que cette erreur sera corrigée dans le prochain numéro de la *Connaissance des Tems* pour 1822.[13]

J'espère aussi que vous me pardonnerez de vous avoir entretenu de choses aussi particulières, et dans une aussi longue lettre; mais un vieillard, qui doit s'occuper assez peu du présent et attend encore moins du futur, vit surtout dans le passé. Dans le cours de mon humble vie, ma principale récompense est la réflexion d'avoir été un zélé promoteur des sciences mathématiques pendant plus de soixante ans, comme professeur principalement à l'Académie royale militaire de Wolwich; connu comme un des plus zélés collaborateurs des Transactions philosophiques, et comme éditeur de cet ouvrage; comme directeur de journaux scientifiques, qui ont été reçus avec de grands encouragemens; et, enfin, comme auteur de différens mémoires, successivement publiés, sur des sujets mathématiques.[14]

Si je ressens cette satisfaction pour une application laborieuse dans la poursuite zélée de ces sciences élémentaires, qui conduisent à la connoissance des principes et des autres ouvrages de notre célèbre Newton, combien doit être agréable pour vous la pensée d'avoir si bien illustré, autant qu'honoré, notre grand auteur, et poussé ses découvertes plus loin que lui-même ne pouvoit peut-être l'espérer.[15]

Que vous jouissiez long-temps de cet avantage, et que vous continuiez à augmenter le faisceau de la science, tel est le vœu de votre obéissant serviteur,

CHARLES HUTTON.

~

Bedford-Row, London, April 9, 1819.

MY LORD MARQUIS,—I have for many years most anxiously wished to have the honour of paying my respects to you in person; but as there is no prospect of my being able to make a journey to France, being now in my 82d year, it only remains to address you in writing, and to request that you will accept the homage of a sincere lover of those sciences which you have contributed so essentially to advance, with the highest honour to yourself, as well as advantage to the scientific world.

An opportunity has just offered for conveying this letter to Paris, and I avail myself of the same to request your acceptance of a copy of my "Tracts on Philosophical and Mathematical Subjects,"[1] which I beg to present to you as a small but sincere token of the high respect which I entertain, in common with the rest of the world, for your genius and talents.

In these Tracts, I wish particularly to call your attention to the leading article in the second volume, "On the Density of the Earth;" and to express my wish that you, who have already effected such profound investigations of its figure, and the doctrine of its tides, would pursue the inquiry respecting its density, so as to correct the errors, or confirm the truth, of my labours. Indeed it seems to me surprising that similar experiments and operations have not been hitherto made in France, where science in every other department has been pursued with the most laudable zeal and the most distinguished success.

As you must naturally feel an interest in whatever relates to this important undertaking, I beg here to state certain circumstances which led to it, with a general account of the manner in which the operations were conducted; and shall conclude by pointing out a slight mistake in the *Connaissance des Tems* for 1821, page 330, line 2,[2] where my labours, in the solution of this important problem, are ascribed to Dr. Maskelyne,[3] a credit which he himself never claimed, but, on the contrary, always corrected the mistake whenever he heard it mentioned.

Some time before the year 1769, when a transit of the planet Venus was expected, the Royal Society contemplated the propriety of sending some skilful astronomers to distant countries, to observe that transit. But as the expense of such a mission was above their finances, they represented their difficulty to the King; when His Majesty was graciously pleased to grant a specific sum to defray the said expense. In consequence, two of my astronomical friends, MM. Wales and Bayly,[4] were in due time sent out to Hudson's Bay, where they passed a summer and a severe winter; and after returning to England, they gave an account of their mission. This was communicated to Government by the Royal Society, together with a statement of the expenses incurred; which being less than the sum that had been granted, they begged to know the King's pleasure touching the surplus; when His Majesty was graciously pleased to reply, that the Royal Society might employ it on any experiment, or to any purpose they might think fit.

After duly contemplating this subject, the Royal Society came to the resolution of employing that surplus in determining, by experiment, the mutual and universal attraction of matter, by observing the deviation of a plumb line from the perpendicular, by the attraction of some mountain; in forming which resolution Dr. Maskelyne took a leading part. In consequence, a correspondence was held with many persons in all parts of Great Britain, to discover a hill or mountain proper for the experiment; which was at length found in the mount Shichallien in the centre of Scotland.[5] The next consideration was, what person was to go to make the experiment? It was much wished that Dr. Maskelyne would go on that mission; but he excused himself, on account that he could not be spared from his duty as Astronomer Royal. After some delay, it was agreed to send a Mr. Burrow, a person who had formerly been Dr. Maskelyne's assistant at the Royal Observatory;[6] who was instructed, first to make an accurate survey of the mountain in all its dimensions, in every direction horizontally, and a great number of vertical sections from all sides; so that from these a perfect model might be made of it, or its magnitude be accurately computed; and then he was to observe the deviation of the *plumb line*. After a year or two had been spent in those measurements, the Royal Society fearing the work might not be well performed, urged Dr. M. to go down himself and direct the operations; which, after obtaining the King's leave, he did accordingly, and brought back Mr. Burrow's survey and measurements, with his own observations of the deviation of the plummet made on both sides of the hill. These operations employed the years 1774, 1775 and 1776; and the method of performing them was stated by Dr. M. in the Philosophical Transactions for the year 1775.[7]

Thus then the problem was performed, so far as showed experimentally that there *is a mutual attraction in all matter;* and that the sum of the two attractions on both sides of the hill in particular, amounted to 11.6 seconds. The Society were very well pleased with the operations. They also conferred on what further use might be made of these measurements; and it was perceived that by their means, after due calculations made, the mean density of the whole earth might be determined: for there were now obtained these data; viz. the mass and attraction of the hill, with the magnitude and attraction of the earth, to determine its density and mass. This was of course a grand and most important object. But the question was, *Who* was to perform the immense calculation that was requisite? who *would do it?* and indeed who *could* do it? There might be one or two persons who were thought capable of undertaking the task; but those and all the other members of the Society stood aghast at the prospect of the number and magnitude of the calculations, and shrank with dread from the mighty labour. I was then looked to, and asked by the President and Council to undertake the task; to which I was also strongly solicited by Dr. Maskelyne.

Being then a *young member of the Society*, and anxious to distinguish myself, I undertook the operation and became responsible for the result; which, after the daily labour of nearly two years, I produced to the Society, with a minute detail of the particulars of the calculations, accompanied with the appropriate drawings. The Society were so well pleased with this work that they awarded me a liberal *premium*, and printed my memoir in the Philosophical Transactions for the year 1778, exactly in the form in which I had delivered it.[8]

It is to be observed, that in this operation, being the first of the kind that was ever made, all the modes of calculation were to be devised or invented; and that, without several plans which I contrived, it might be doubted whether the labour could ever have been accurately accomplished. To the ingenuity of these contrivances the most honourable testimony has lately been borne by one of the best mathematicians in this kingdom, Professor Playfair.[9] The result of that calculation was, that the mean density of the earth is to that of the hill, in the ratio of 9 to 5. What was the density of the hill, was then not known, and therefore the density of the earth, as compared to water, could not then be determined. But to show the mode of making the comparison, I assumed for the sake of example, the smallest density of stone, $2\frac{1}{2}$, though this was obviously below the real density of the rock of the hill; then $\frac{9}{5} \times \frac{5}{2} = \frac{9}{2} = 4\frac{1}{2}$, would be less than the density of the earth, as the numbers to be adopted till we could know the real density of the hill.

Many years after this, a lithological survey was made of the same hill by Professor Playfair, who found it to consist of rocks of a great variety of densities from 2.5 to 3.2, the medium being about $2\frac{3}{4}$ or 2.8. Assuming the former number then $2\frac{3}{4}$, as the density of the hill; then $\frac{9}{5} \times 2\frac{3}{4} = \frac{99}{20} = 4\frac{19}{20}$, or nearly 5, results for the mean density of the earth. And this result, instead of the former, I substituted in its proper place in my abridged edition of the Philosophical Transactions, from the beginning, to the end of 1800, in eighteen large quarto volumes, where it may be seen.[10] If we assume 2.8, instead of the $2\frac{3}{4}$ above used, then $\frac{9}{5} \times 2.8. = 5.04$ comes out for the density of the earth, a little above 5, as the former was a little below it. I fixed therefore on the number 5, as my result for the mean density of the earth, determined by my calculation. The same conclusion is also given in my Tracts in the first memoir of the second volume, where the whole process is stated at large,[11] as may be seen in the set of those Tracts now presented along with this letter; and as they were also formerly presented to the Royal Institute.[12]

And now, after all these evidences of my being the first, and only person, who had calculated the mean density of the earth from the original measurements, and have been considered and acknowledged as such for almost half a century, the world must be astonished to perceive the honour of that determination in danger of being transferred to another person who never expected nor desired it, unless the accidental mistake or misnomer in the *Connaissance des Tems* be acknowledged and corrected. I hope and trust therefore, from the known honour and liberality of your character, that the mistake will be corrected in the next Number in the *Connaissance des Tems* for 1822.[16]

I hope you will pardon my troubling you with this long letter, and so many minute particulars. But old men, who can enjoy but little of the present, and expect still less from the future, live chiefly in retrospect. In my humble walk in life, my chief gratification is the reflection of having been a zealous promoter of mathematical science for upwards of sixty years, as a teacher, chiefly at the Royal Military Academy at Woolwich; as a laborious contributor to the Philosophical Transactions, and an Editor of that work; as the conductor of certain scientific Journals which have met with great encouragement; and also as the author of several successful publications on mathematical subjects.[14]

If I feel this satisfaction in a retrospect of laborious application in the zealous pursuit of those elementary sciences, which lead to the knowledge of the Principia and other works of our immortal Newton; how much more gratifying must your reflections be, in having so ably illustrated as well as honoured our great author, and carried his discoveries higher than perhaps he himself could have ever contemplated![15]

That you may long enjoy this gratification, and continue to enlarge the boundaries of science, is the sincere wish of

Your most obedient and faithful servant,

CHARLES HUTTON.

To the Marquis de Laplace.

Notes

Pierre-Simon, Marquis de Laplace (1749–1827), mathematician, astronomer, and physicist, was by this date well known in Britain as the author of the *Traité de Mécanique Céleste*, the appearance of which from 1799 onwards had provoked comment on the state of British natural philosophy including by both John Playfair and Robert Woodhouse (see notes on Letters 19 and 85): DSB; see also Letter 122.

I take the sequence of events to be as follows. The French version of the letter was presumably that sent to Laplace in April 1819 (the most natural assumption is that

Hutton wrote it himself in that language; the density of—mostly fairly mild—anglicisms is comparable to that in Letter 24, which is certainly of his own composition). Receiving no reply after several months, he then evidently conveyed the letter to the *Philosophical Magazine*; the English translation printed there was slightly more cautious in its claims about certain of Hutton's publications (see notes 11 and 14) and was therefore perhaps his own work too. Next, perhaps in response to the letter's appearance in print, the editor of the *Journal de physique* obtained the French text and printed it. Finally, in September 1820, Laplace replied (Letter 122).

[1]Hutton, *Tracts* (1812).

[2]de Laplace, 'Sur la figure de la Terre', 330: 'les résultats que Maskeline et Cavendish ont obtenus ...'.

[3]For Nevil Maskelyne see note to Letter 13.

[4]For William Wales see note to Letter; 4. William Bayly (1738–1810), astronomer, also sailed on Cook's second and third voyages of exploration in the 1770s; he subsequently worked as a teacher: ODNB.

[5]Spelt *Schehallien* in Maskelyne, 'An Account' and Hutton, 'An Account of the Calculations'; now more usually *Schiehallion.*

[6]For Reuben Burrow see note to Letter 4.

[7]Maskelyne, 'An Account'.

[8]Hutton, 'An Account of the Calculations'. The minutes of the Council of the Royal Society confirm that Hutton was given the sum of £150 for his work on the calculations (Royal Society, CMO/6, p. 333: Council meeting of 18 June 1778); his designation of the money here as a 'liberal premium' (and the index to the Council minutes—CMO/8A, *s.v.* Hutton, Mr—refers to the payment as 'a Gratuity') might be compared with his remarks in the postscript to Letter 67 about pay for advice on a bridge project.

[9]Reference here and below was to Playfair, 'An Account of the Lithological survey of Schehallion'.

[10]Hutton, 'An Account of the Calculations' was abridged at Hutton, Shaw, and Pearson, *The Philosophical Transactions ... abridged*, vol. 14, pp. 408–23. At p. 420 Playfair's lithological survey was mentioned and final result was consequently amended to a ratio of five to one, without a clear indication that this matter was not in the original paper.

[11]The first tract in the second volume of Hutton's *Tracts* (1812) was in fact a reprint of Hutton, 'An Account of the Calculations', with modifications towards the end as indicated in the English text of this letter.

[12]That is, what is now the Institut de France; from 1814 to 1848 it was known as the Institut Royal de France.

[13]Le nouveau volume de la *Connaissance des Tems* vient d'être publiée, mais sans qu'il y soit question de cette erreur; et même dans un nouveau Mémoire, semblable au premier, non-seulement la mention du véritable nom et de l'expérience est omise, mais on fait une recommandation particulière d'une autre expérience exécutée très-postérieurement, par Cavendish, avec un appareil qui avoit été préparé par l'inventeur de cette expérience, le rév. M. Michel.—Fév. 1820. [*Note in source.*] For Cavendish's 1798 experiment see note to Letter 81.

[14]Some of the claims in this sentence—which were perhaps slightly weaker in the English version of the text—can scarcely be sustained. As a member of the Council of the Royal Society in 1779 and 1780, Hutton was *ex officio* a member of the committee of papers which selected material for printing in the *Philosophical Transactions*, and he was for two years responsible for translating the foreign papers which were printed in it; but he was in no other sense an 'editor' of that journal. The only other periodicals Hutton is

known to have 'conducted' were certain annual almanacs and the *Ladies' Diary*, and it was stretching a point to describe these as 'scientific Journals' even in the usage of the period.

[15]The allusion was to De Laplace, *Mécanique céleste*, which had received much notice in Britain.

[16]That Number of the *Connaissance des Tems* has been lately published, but without any direct notice of the mistake above alluded to; and even in a new memoir, similar to the former, not only is the mention of the very name and experiment omitted, but special commendation is bestowed on another far inferior experiment by Mr. Cavendish, performed by an apparatus, which was prepared by the inventor of that experiment, the Rev. Mr. Michell.—Feb. 1820. [*Note in source.*]

Letter 121.
CHARLES HUTTON to CATHERINE HUTTON, 18 October 1819

Birmingham, Archives, Heritage and Photography Service, MS 3597/73.
In the hand of Isabella Hutton except for the signature.

Bedford Row
October 18th. 1819.

My Dear Cousin,

I am grateful for the sentiment of your apology for not writing, but am sorry that your humanity should have so long prevented us from the opportunity of rejoicing at the improvement of your health. I say we, because, as you see, this epistle exhibits the conjoint handywork and effusion of our affections & congratulations. We are glad to learn, not only the improvement in your health and strength, but also the manner of it, partly by the means of both milk and meat, because it is exactly my own regimen, & by which & the good nursing of your sister[1] I am greatly improved also. So that I heartily join in your opinion to repel the unnatural regimen of our friend Sir Richard,[2] but must dissent from his new philosophy also.

So you made good your journey to the West, & hope to your amusement & benefit of your health, though not by means of your warranted best horse in the world. Are glad you took your revenge of the natural brute, & brought him to his stumps. But am sorry you did not get the same of the human culprit who warranted for him, & against whom you doubtless had a legal as well as a just right of revenge.

But, instead of your excursion westward, we fully expected to learn the account of your perilous one towards the arctic regions. For about some two months ago, in a letter from my son's family[3] who have been some time at Aberdeen, it is stated that the Postman in that town brought to them a letter addressed to Miss Catherine Hutton at Aberdeen, thinking they might know the lady, as he could find no such person in the place. How is this to be explained? There and thitherward we thought you must see plenty of new grounds you had never seen before.

As for myself, tho' much better than when favored with your sight, I am no Sir Isaac Heard,[4] and despair of ever being able to scamper over the hills again, whatever be my inclination. Apropos, now we speak of journeying, I think we had a kind of promise to have the pleasure of seeing our other Cousin your brother[5] here in the course of this year. This disappointment is the more to be regretted, as I hoped by his means to have sent you a beautiful engraved Portrait of our other mutual friend and cousin James Hutton,[6] which I have procured for you. Apropos again, have you seen the new work of Dr Franklin's letters?[7] if not, you must at any rate see the first vol. where, at pa. 113, you have a letter to our cousin Ja^{s}. Hutton; & such a letter!!!

In compliance with my dear cousin's wish, I have been getting new portraits of my phiz, with a view to a new engraving, both in my Cap, & without it;[8] but am cruelly disappointed. I mean however to make one more attempt for it.

With regard to the questions of our friend Mr Benj. Bright,[9] I am not aware of any other authority for the particulars of the life & transactions of Robert Hues than what are mentioned in my Dictionary.[10] Of his two works there noticed, I possessed the one viz. the De Globis Cœlest. et Terrest. Tract. 2. but the other, the Breviarium Totius Orbis,[11] I never possessed, and indeed never saw, nor do I believe there exists a copy of it in all this great metropolis, at least there is not one in the British Museum, where I have searched in vain; probably a copy might be seen in Oxford, where Hues was educated and spent many years of his life.

Now that you are settled at home, and your health improved, I hope Africa will make a fresh and rapid progress;[12] for which you will doubtless see the new work of Mr Bowditch on his visit to the kingdom of Ashantee.[13]

In the best wishes for your happiness in that and all your other concerns your Sister unites in prayer with,

My Dear Cousin
Yours affectionately
Chas. Hutton

Miss Hutton
Bennetts Hill
Near Birmingham

Notes

For Catherine Hutton see note to Letter 119.

[1]That is, Isabella Hutton (see note to Letter 16).

[2]Possibly the author and publisher Sir Richard Phillips (1767–1840), who was both an advocate of strict vegetarianism and an author of eccentric philosophical and natural philosophical work: ODNB. He would have been known to Hutton as the founder of the *Monthly Magazine*, in which Hutton published a letter in 1802.

[3]For Henry Hutton see note to Letter 12.

[4]The herald and, from 1784, Garter King of Arms Sir Isaac Heard (1730–1822) was by this date notable for his vigorous health despite his advanced age: ODNB.

[5]Thomas Hutton (1757–1845): Jewitt, 'Pedigree of the family of Hutton, of Derby, Birmingham'.

[6]This reference was to the Moravian minister James Hutton (1715–95: ODNB). Nineteenth-century reports, including an obituary quoting words spoken by Charles Hutton, stated that Hutton's mother and the mother of James Hutton were sisters. (Obituary of Hutton in *The Literary Gazette* (1 February 1823), pp. 75–6; letter of Charles Blacker Vignoles to the Marquis of Northampton, 25 March 1839 (Portsmouth, History Centre, Vignoles Letters, Letter 751); Weld, *Catalogue*, 52; cf. Galton, *Hereditary Genius*, 220–1 and K.H. Vignoles, *Romantic Engineer*, xii: not all had exactly the same version of the relationship, and Galton (p. 221) seems to have assumed that Dr James Hutton (the geologist) was involved, whereas the story in its earliest recorded form (*The Literary Gazette*) made its reference to the Moravian unambiguous by a description of Cosway's portrait of him.)

These sources also stated that James Hutton's grandmother was a sister of Isaac Newton's mother, but as Galton pointed out, this was barely possible chronologically. Sources closer to James Hutton, including a piece in the *Annual Register* probably written by him, stated instead that his mother was the granddaughter of James Ayscough, whose sister Hannah was Newton's mother. ('I.H.', Letter to the editors, *Monthly review* (1772), 36–7; 'J.H.', 'New Anecdotes of Sir Isaac Newton', *Annual Register* (1776), 24–7. Brewster, *Isaac Newton*, 290 stated that the author of the latter was James Hutton; see also p. 316. Similar reports appeared in Benham, *James Hutton*, 8, 9, and ODNB, both apparently based on the same sources.) There seems no particular reason to doubt the relationship of James Hutton to Newton in this form, for which the earliest printed source in 1772 seems to have been based on notes made by members of the previous generation, who had known Newton. That of Charles to James Hutton seems more problematic, since it involves the coincidence of two sisters marrying men with the same surname (Hutton); nor have I been able to locate an Eleanor Ayscough (Hutton's mother, if the relationship to James Hutton is correct) in the genealogical records. But there seems nevertheless no very clear reason to reject it, and in its favour is the fact that Lord Stanhope, who apparently knew the family of James Hutton, left the famous Vanderbank portrait of Newton to Hutton in his will. (Brigden, 'Samuel Wesley, Junior, and His Circle, 1690–1739 [part III]', 99–100; also Public Record Office, PROB 11/1590/64: Will of Charles Earl Stanhope.)

[7]That is, Franklin and Franklin, *Correspondence of Benjamin Franklin*, which was in fact in its third edition by the time this was written.

[8]At least three portraits of Hutton were made during this late period of his life, by Philip Reinagle, Andrew Morton, and William Read; the two last certainly resulted in engravings and the Read portrait could well be called a cruel disappointment: but it is not really clear whether Hutton referred to any of them here. See (1) K.H. Vignoles, *Romantic Engineer*, 4; (2) Watson, *Literary and Philosophical Society*, 100, also reproduced in ODNB and www.bbc.co.uk/arts/yourpaintings/paintings/charles-hutton-17371823-lld-frs-mathematician-57703; and (3) Wellcome Library, Iconographic Collections, ICV No 3190. See also Letter 119.

[9]This seems likely to have been the Benjamin Haywood Bright whose large collections of manuscripts and books were sold in 1844 and 1845 respectively: Bright, *Catalogue ... of manuscripts*, Bright, *Catalogue of the ... library*.

[10]Robert Hues (1553–1632), mathematician and geographer, was discussed in Hutton's *Dictionary* (1815), vol. 1, pp. 657–8 (he did not appear in the first edition).

[11]The first work mentioned was Hues, *De globis*. The second seems to have arisen from a mistake by Bright, who had probably seen a copy or a description of Hues, *Tractatus duo mathematici*: this was a reprint of Hues, *De globis* issued together with the *Breviarium totius orbis terrarum, una cum serie Romanorum Imperatorum* of Petrus Bertius (1565–1629).

[12]Referring to Hutton, *Africa*: see Letter 119. Two more volumes were published in 1821.

[13]Bowdich, *Mission*.

Letter 122.
PIERRE-SIMON DE LAPLACE to CHARLES HUTTON, 11 September 1820

Printed (in English) in the *Philosophical Magazine*, ser. 1, 56 (November 1820), 321–2; also reprinted verbatim as a separate pamphlet.
No French printing known.

Paris, Sept. 11, 1820.

SIR,—I hope you will have the goodness to accept my excuse for not replying sooner to the letter with which you honoured me. I expected every day that the Tracts I am printing in the volumes of the *Connoissance des Tems* would give me an opportunity of inserting what you seemed to wish. But as this opportunity has not yet occurred, I have resolved to publish, for the express purpose, a Memoir on the Density of the Earth, in the volume of the *Connoissance des Tems* for 1823, which will soon appear. I have the honour to send you a copy of this paper. It is my wish that it may satisfy you; being very desirous to prove to you how much I esteem your talents, and honour your person. I thank you for the present you have been so good as to send me of the collection of your Tracts.[1] I have been long acquainted with your profound researches, which secure to you a distinguished rank among geometricians, and which have long inspired me with a high esteem for you, the expression of which I beg you will accept.

M. DE LAPLACE.[2]

Notes

For Pierre-Simon, Marquis de Laplace see note to Letter 120, to which this was the reply.

[1]Hutton, *Tracts* (1812).

[2]In both English printings of this letter there followed an English translation of Laplace's short treatise 'Sur la densité moyenne de la terre', which appeared both in the *Connaissance des tems ... pour l'an 1823* (Paris, 1820), 328–31 and in the *Journal de Physique, de chimie, et de l'Histoire Naturelle* 91 (1821), 146–50. It used the information from Hutton's letter (Letter 120) and gave proper credit to Maskelyne, Hutton, Playfair, and Michell for their work on the subject.

Letter 123.
CHARLES HUTTON to ALEXANDER TILLOCH, 23 October 1820

Edinburgh, University Library, La. II. 425/25.
In the hand of Isabella Hutton.

34 Bedford Row
Octr. 23. 1820.

Dear Sir,

Will you have the goodness to favor me with a call some early day this week, as you pass by to Islington, that I may put into your hands a most curious and important article for your Magazine, being an interesting communication from M. de Laplace, consisting of a kind & friendly letter in answer to mine that was formerly sent to him, & was inserted in the Magazine of Feb. last,[1] which he has also accompanied with a new and highly interesting Memoir on the subject of the Earth's Density &c.[2] I am

Dear Sir,
Yours very truly
Chas. Hutton

Alex^{r}. Tilloch Esqr.
Picket Place
Temple Bar

Notes

The journalist and inventor Alexander Tilloch (1759–1825) was in 1798 the founder of the *Philosophical Magazine*, in which Hutton had published articles in 1811, 1820, and 1821 and in which Letter 120 had appeared earlier in 1820: ODNB; see also Letter 124.

[1]That is, Letter 120, which had appeared in the *Philosophical Magazine* in February 1820.

[2]The letter referred to was Letter 122; the memoir Laplace, 'Sur la densité moyenne de la terre'; both were indeed subsequently printed in the *Philosophical Magazine* in November 1820.

Letter 124.
CHARLES HUTTON to ALEXANDER TILLOCH, October–November 1820

Printed as an initial note to Letter 122 in the *Philosophical Magazine*, ser. 1, 56 (November 1820), 321; also printed as an initial note to the separate reprinting of that letter.

SIR,—IN the Philosophical Magazine of February last, you were so obliging as to insert a note from me on the subject of my Calculations on the Density of the Earth; for I then apprehended that the credit of that operation had been transferred to others: and I scarcely need observe, that as honour is the chief reward of scientific labours, it was natural that I should feel some uneasiness on being unjustly deprived of what I conceived my right.

You also had the goodness at the same time to insert the copy of a letter on the same subject, which I had addressed to the *Marquis De Laplace*,[1] under an idea that his Lordship had forgotten to acknowledge the receipt of that little communication. In this apprehension, however, I am happy to find that I was mistaken, as the delay arose, not from any neglect on his part, but rather from his attention to the subject, in preparing a profound and interesting paper on the earth's density, a copy of which he has transmitted to me, with the following very kind and obliging letter.[2]

To Mr. Tilloch

Notes

For Alexander Tilloch see note to Letter 123.

[1]That is, Letter 120.

[2]That is, Letter 122 and the treatise 'Sur la densité moyenne de la terre'.

Letter 125.
CHARLES HUTTON to FRANCIS BAILY, 17 September 1821

London, Senate House Library, MS 913B/3/3 (xv): ALS.

Sep. 17, 1821.

My Dear Sir,

I have to thank you very much for the present of your useful paper on the fixing of a Transit Instrument, &c.[1]

I have also, in the name of the Committee who have composed and issued the papers proposing a bust of your hble Servant,[2] to intreat your forgiveness for the liberty they have taken by inserting your name in the accompanying papers, which they deemed essential to its success, without your leave, which they were not able to procure on account of your absence from town, and which they only risked on my venturing to take the blame on myself.

With entreaties of your kind indulgence as above, I remain, with great regard

Yours very truly
Chas. Hutton.

Francis Baily, Esq.

Notes

For Francis Baily see note to Letter 102.

[1]Baily, 'Method'.

[2]During 1821 a printed prospectus was circulated under the title 'Subscription for a marble bust' and notices were placed in periodicals including the *Gentleman's Magazine* (November 1821, p. 452). Those named as involved in the project were ('Subscription',

p. 2) James Andrew (1774?–1833, headmaster of the East India Company's military college at Addiscombe: ODNB), Francis Baily, Olinthus Gregory, Patrick Kelly (see notes to Letters 102, 118, and 108 respectively), Daniel Moore (probably the F.R.S., *c.* 1760–1828: 'List of Fellows of the Royal Society 1660–2007' (2007), available from www.royalsoc.ac.uk/library), and Edward Troughton (1753–1835, maker of scientific instruments: ODNB). Catherine Hutton in Birmingham and William Armstrong in Newcastle appear also to have been involved in collecting subscriptions: see *Tribute of Respect*, 3, and Letter 130.

Letter 126.
Catherine Hutton to Charles Hutton, 4 November 1821

Printed in Hutton, *Reminiscences*, 180.

Bennett's Hill [near Birmingham],
Nov. 4, 1821.

My dearest Cousin,—

I received your letter with great pleasure, and read it with greater. Such a proof of health and vivacity quite astonished me! There is not one thought or line which savours of old age. Never mind your saint, you are a match for all the saints in the calendar.[1]

I am glad you will leave to posterity so durable a resemblance of your countenance as a marble bust;[2] marble may last as long as mathematics; while painting, even that of Sir Isaac Newton, must perish. My brother[3] begs you will bespeak a cast for him; and, in its features, I shall contemplate with pleasure, not the philosopher who will interest all men of science, but my honoured cousin and kind friend. You will do us the favour to enter my name, and that of my cousin, Samuel Hutton, as subscribers.

To Dr. Hutton, 34, Bedford Row.

Notes

For Catherine Hutton see note to Letter 119.

[1]Catherine Hutton was evidently responding to a lost letter. Hutton was born on 14 August and baptized on 4 September, neither of which was the feast day of any saint in the Anglican calendar of this period; it is therefore not clear whom he regarded as 'his' saint.

[2]See note to Letter 125.

[3]Thomas Hutton: see note to Letter 121.

Letter 127.
PATRICK KELLY to CHARLES HUTTON, 24 December 1821

London, Wellcome Collection, MS 5270 no. 70: fragment.
A sheet trimmed (with loss of text) and pasted to a guard sheet; the cover is lost. On the guard sheet appears a remark in the hand of Catherine Hutton (for whom see note to Letter 119):

> I knew D^{r}. Patrick Kelly, and I knew no good of him. The above was addressed to his friend and mine, D^{r}. Hutton[.] The work of mine alluded to was never finished.

Finsbury Lyne
Dec. 24. 1821

My Dear Sir

I have perused Miss Huttons Sketch of ⟨"The⟩ Female History of England" and I really think it a work which when completed according to the specimen must spread extensively.

[*Text lost*]

It may be worth her while to mentio[n] the great Sanctity & purity of the Female Character — even in the Dar[k] Ages of superstition. — a period which occupies a large portion of History. — The Lives of the Sain[ts] will afford abundant information

[*Remainder of text lost.*]

Notes

For Patrick Kelly see note to Letter 108.

Letter 128.
CATHERINE HUTTON to CHARLES HUTTON, 11 February 1822

Extract printed in Hutton, *Reminiscences*, 180–1.

Feb. 11, 1822.

One of my desultory occupations has been to read 'the Pirate' [Sir Walter Scott's],[1] and this gave me some consequence when I went to Cheltenham, as

I was the first person at Captain Tudor's table who had read it, and everyone was anxious to hear of it. When other persons were as well informed as myself they thought less favourably of it than I did. I like the hitherto unknown scenery of the islands, the manners of the islanders, the old Udaller and his daughters, and even the Pirate, to whom I would have given a happier lot at last. If I must have a witch, I like Norma, because her exploits and predictions are accounted for upon rational principles, a circumstance in which the other fanciful characters of this author are woefully deficient.

Note

For Catherine Hutton see note to Letter 119.

[1]Scott, *The Pirate*.

Letter 129.
CHARLES HUTTON to JOHN BRUCE, 22 March 1822

Printed in Bruce, *Memoir*, 39–42.

Bedford Row, 22*d March*, 1822.

MY DEAR SIR,

I am ashamed so long to have delayed acknowledging your kind and excellent letter of Feb. 23, a delay for which your goodness will find an excuse and forgiveness in my unfitness, and consequent reluctance, for writing. For the valuable contents of your letter I must thank you very much indeed, as well as for the justice you do me in supposing I am greatly interested in the welfare and honour of every thing that concerns my native place; and particularly the few worthy characters whom I have the pleasure to call my personal friends. I am glad the Literary and Philosophical Society has at length fixed on a site for their new and permanent situation, and that so central and convenient, and so near the last abode of myself.[1] I know the house well, and have been often in it; and which, when I left Newcastle, was, and had been for many years before, the residence of Mr. Gibson, the Town Clerk, who was much of a gentleman, and every way a worthy character.—The house was always esteemed a very excellent one, with a fine open area in front, and spacious garden behind, extending to the Town Walls, making a conspicuous figure in my own engraved plate of Newcastle.[2] I trust that your Building Committee will well deliberate and select the plan for the new buildings, respecting rather usefulness and economy, than mere show and elegance.

When agreed on, I shall be grateful for a small copy of the plan and elevation. Indeed I will venture to hint the propriety of having executed, for general use and sale, a small engraved view of the whole, either for general sale, or for insertion in some magazine, accompanied with a description of that work and the Society, which may probably attract additional subscriptions to ten times the amount of the expense of the publication.

When now speaking of expense, I beg of you to announce my subscription of £20. towards the new building, which I shall transmit through our friend Mr. Riddle,[3] whom I have the pleasure of seeing now and then, and who seems to be happy in his new situation at Greenwich, where his labours will doubtless be very valuable and useful.

On the subject of the final future destination of the marble bust,[4] I observe your delicate hints, which I assure you are much in unison with my own views and wishes, though it would, perhaps, be improper to make an immediate premature promise. Would to God, also, that you had given the hint in time about my books, which might have deprived my enemy of his triumph in disappointing me, and have secured to your Society the satisfaction and credit of possessing the best mathematical library in the kingdom.[5] But lamenting now is useless, though we now hear frequently the execrations against the late President of the Royal Society of London,[6] for having caused the *dispersion* of the only library that future mathematicians could have had recourse to.

Yes, dear Sir, I observed the deep affliction that our friend Mr. Turner must have felt for the death of his excellent son,[7] of whom I perceive some account, perhaps by the parent himself,[8] in the last number of the Newcastle Magazine. I had not courage to write to him on the occasion, but wish you to mention my respects and condolence to him.

I am glad to have the pleasure to inform you, that my health has much improved, and that it appears to have now no very great enemies but damp, uncomfortable weather and old age, against both which I fear there is no protection. But so much better do I feel now that, if it continue, I do not despair of being able to follow my wishes to have the pleasure to run down when your new buildings are completed,[9] to have the pleasure of seeing them, and paying my grateful respects to my present remaining friends in Newcastle.

With my daughter's respects[10] and my own,

I am, dear Sir,

Your obliged and affectionate servant,

CHAS. HUTTON.

To Mr. Bruce.

Notes

For John Bruce see note to Letter 104.

[1]For the Newcastle Literary and Philosophical Society see note to Letter 70. The new building referred to was on Westgate Street, on which Hutton's purpose-built school house also stood, occupied by him until 1773.

[2]Hutton, *Plan*.

[3]Edward Riddle (1786–1854), mathematician and astronomer, taught mathematics in the north-east of England, was noticed by Hutton as a distinguished contributor to the *Ladies Diary*, and at Hutton's recommendation became successively master of the Trinity House School at Newcastle and master of the upper mathematical school at the Royal Naval Hospital, Greenwich: ODNB.

[4]See note to Letter 125; Bruce (p. 39) had suggested that Hutton leave the bust to the Newcastle Literary and Philosophical Society. In his will, dated 7 January 1823, he did so; it was the only item from his estate not left to his daughter Isabella: the National Archives, Kew, PROB 11/1666/375, fol. 297^{v}, will of Charles Hutton.

[5]See Letter 115; Bruce (p. 39) had expressed regret that he 'had not in time given him a hint that might have saved the dispersion' of Hutton's library.

[6]For Joseph Banks see note to Letter 22.

[7]The Rev. Henry Turner, Minister of the High Pavement, Nottingham. [*Note in source.*] For William Turner, the father, see note to Letter 104.

[8]By the Rev. J.G. Robberds, of Manchester. [*Note in source.*]

[9]On Hutton's projected visit to Newcastle see also Letters 115, 113, and 130.

[10]For Isabella Hutton see note to Letter 16.

Letter 130. CHARLES HUTTON to WILLIAM ARMSTRONG, 20 April 1822

Printed in Bruce, *Memoir*, 42–3.

Bedford Row, 20*th April*, 1822.

MY DEAR SIR,

I am just favoured with your kind letter, so very honourable by its contents of respects from my dear and most respectable townsmen;[1] the recollection of whom is at all times so enlivening to my heart. Pray, dear Sir, have the goodness to present the grateful respects of a humble individual to all his too partial townsmen, for their kind and consoling testimonials of the recollections so consoling and flattering to his breast. Particularly the Worshipful Corporation, the Brethren of the Trinity House, the Gentlemen of the Exchange, those of the Literary and Philosophical Society, &c., and say that if health permit me, I fully intend to do myself the honour and great pleasure of accomplishing in summer my pilgrimage to the dear place of my nativity; though ill able now to visit and revise the many places of my youthful scenes of infantile delights.[2] And thank you, dear Sir, for your kind advice and offer of assistance of accommodation in the conveyance, which will be very acceptable to so ancient and so novel a traveller. I was also, the day after you parted from me, honoured with a visit from Sir John Swinburne, the worthy president of your Literary and Philosophical Society,[3] and had the pleasure of a long and interesting conversation on scientific and intellectual subjects. He also was so kind as to solicit my visit to Newcastle, and kindly recommended the same easy mode of conveyance that you mention, and which indeed is the only one which I could venture to encounter; for I well remember that in my youth, more than half a century ago, I was almost shaken to pieces by land conveyance, in two or three journeys then made to London.[4]

I am, dear Sir,
Your much obliged,
CHAS. HUTTON.

To Wm. Armstrong, Esq.

Notes

William Armstrong (1778–1857), corn merchant and local politician, of Tyneside, was a contributor to mathematical periodicals; he was a supporter of the Newcastle Literary and Philosophical Society and was involved with the project for its new building mentioned in Letter 129: ODNB; Welford, *Men of Mark*, vol. 1, pp. 106–8; also Wallis, 'Newcastle mathematical libraries', 1–9.

[1]Mr A. had communicated to the Doctor information of various subscriptions in this town, towards defraying the expenses of the bust. [*Note in source.*] For the bust of Hutton see note to Letter 125.

[2]On Hutton's projected visit to Newcastle see also Letters 115, 113, and 129. This sentence hints at a desire to conceal the details of Hutton's childhood in the pits and pit villages of north Tyneside.

[3]Sir John Edward Swinburne, Bt. (1762–1860) was President of the Newcastle Literary and Philosophical society from 1798 to 1838: Watson, *Literary and Philosophical Society*, 105; Welford, *Men of Mark*, vol. 3, pp. 476–8.

[4]It was reported in a letter of S. Barrass to Thomas Wilson in 1825 (Woodhorn, SANT/BEQ/26/1/8/584) that Hutton visited London on business connected with his 1770 *Mensuration*, and that it was on such a visit that he was made aware of the vacancy at Woolwich. He also travelled to London for the examination for that position; and a third time when actually relocating to Woolwich. It may be implied here that these were his only journeys to London from Newcastle.

Letter 131. CATHERINE HUTTON to CHARLES HUTTON, 10 July 1822

Extract printed in Hutton, *Reminiscences*, 181.

Bennett's Hill [near Birmingham],
July 10, 1822.

My dear and much respected Cousin,—

...I have written to Mr. Hamper,[1] telling him that, from the similarity of their tastes and pursuits, I had mentioned him to General Hutton,[2] who would be glad to see him when he went to town. He sent me a very pretty note in return, which the General shall see when opportunity offers, saying he shall be happy to call on General Hutton, or to see him at Deritend House if he comes into Warwickshire. In this latter case he invites me to be the General's aide-de-camp. Such a mouse as I aide-de-camp to a General officer! It is true, the only enemies to be attacked are antiquarian spiders; and these Mr. Hamper promises to dislodge before his visitors arrive; but Mistress Mouse could as soon face an army as examine Mr. Hamper's antiquities.

CATHERINE HUTTON.

To Dr. Hutton, Bedford Row, London.

Notes

For Catherine Hutton see note to Letter 119.

[1]Mr Hamper was an eminent Birmingham antiquary. He resided in Moseley Street, at what was formerly known as the Apollo House, an hotel with beautiful gardens, commanding an uninterrupted view of Bromsgrove Lickey. Mr Hamper converted this house into a private one, and re-named it Deritend House. [*Note in source.*]

[2]For Henry Hutton see note to Letter 12.

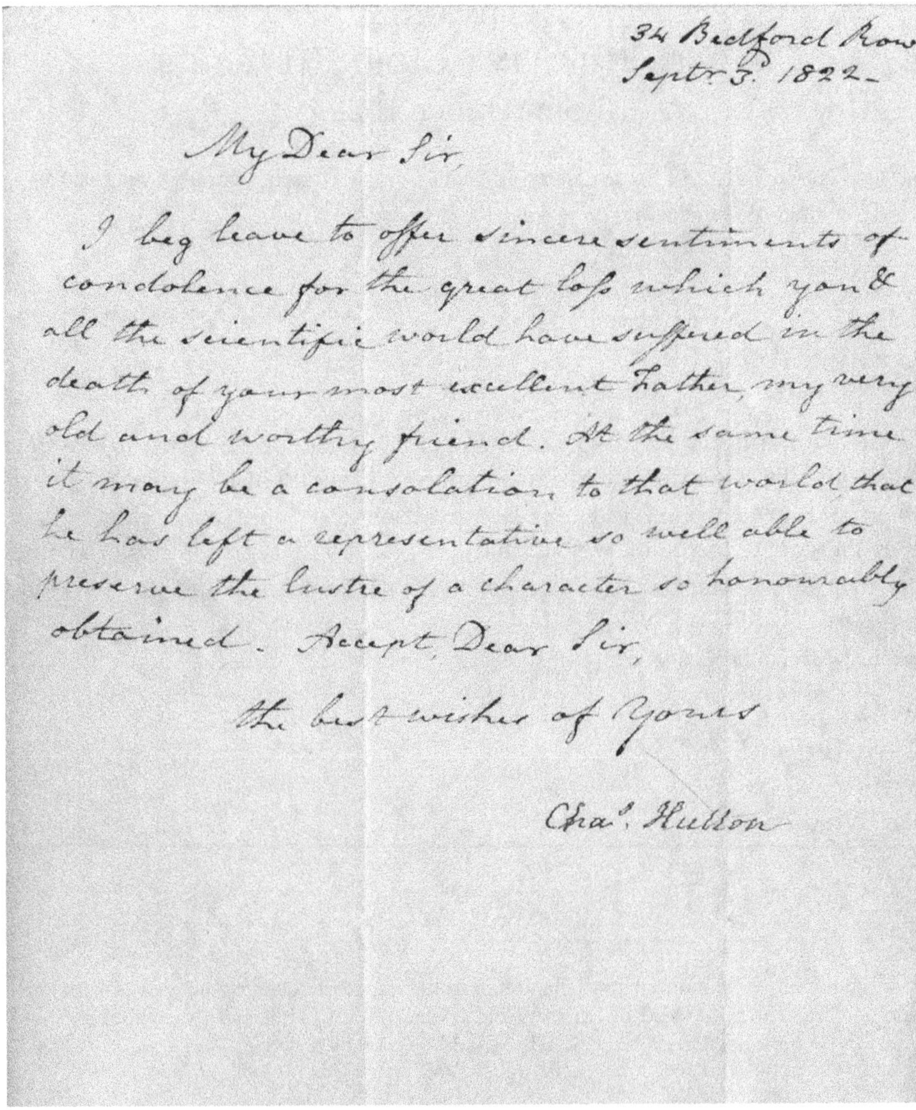

34 Bedford Row
Septr. 3d. 1822-

My Dear Sir,

I beg leave to offer sincere sentiments of condolence for the great loss which you & all the scientific world have suffered in the death of your most excellent Father, my very old and worthy friend. At the same time it may be a consolation to that world that he has left a representative so well able to preserve the lustre of a character so honourably obtained. Accept, Dear Sir,

the best wishes of Yours

Chas. Hutton

Figure 5: Letter 132: Charles Hutton to John Herschel, 3 September 1822 (in the hand of Isabella Hutton). Austin, TX, University of Texas, Harry Ransom Center, Herschel Family Papers 30.19. Reproduced by kind permission of the Harry Ransom Center, University of Texas at Austin.

Letter 132.
CHARLES HUTTON to JOHN HERSCHEL, 3 September 1822

Austin, TX, University of Texas, Harry Ransom Center, Herschel Family Papers 30.19.
In the hand of Isabella Hutton except for the signature.

34 Bedford Row
Septr. 3^{d}. 1822.

My Dear Sir,

I beg leave to offer sincere sentiments of condolence for the great loss which you & all the scientific world have suffered in the death of your most excellent Father, my very old and worthy friend. At the same time it may be a consolation to that world, that he has left a representative so well able to preserve the lustre of a character so honourably obtained. Accept, Dear Sir,

the best wishes of yours
Chas. Hutton

J.F.W. Herschel Esqr.
Slough
Near Windsor

Note

Sir John Frederick William Herschel, first baronet (1792–1871), mathematician and astronomer, was the only child of Hutton's old friend William Herschel (for whom see note to Letter 15; also Letters 25, 26, 27, 28, 29, and 55): ODNB, DSB.

Letter 133.
JOHN SCOTT, EARL of ELDON, to HENRY HUTTON, 3 February 1823

Printed in Bruce, *Memoir*, 47 (our source); also in obituaries of Hutton in the *European Magazine*, *Gentleman's Magazine*, *Monthly Magazine*, *Newcastle Magazine*, and a number of other later accounts of Hutton's life.

Feb. 3d, 1823.

SIR,

I request you to accept my very sincere thanks for your communication received on Saturday last.[1]

Full sixty years have passed since I had the benefit of your venerable father's instructions, and that benefit I regard as one of the many blessings which I have enjoyed in life, and of which blessings I wish I had been more worthy.

I feel very painfully that I did not wait upon Dr. Hutton personally to thank him for his letter, in which he wrote with such remarkable and affecting kindness respecting Lady Eldon and myself, both his pupils.[2] I shall preserve that letter as a testimony, that a person of his eminence had, through so many years, recollected us with a sort of parental affection.[3]

I shall not fail to preserve anxiously the medal which you have been pleased to send to me, and for which I beg you to receive my thanks.[4] To secure to his memory the respect and veneration of his country, this memorial was not wanting; he will long be remembered by a country so essentially benefited by his life and works.

I am, Sir,
Your obedient and obliged servant,
ELDON.

Notes

John Scott (1751–1838), the son of a Newcastle coal factor, had been sent to Hutton for mathematical tuition while he was a pupil at the Royal Free Grammar School in Newcastle, probably around 1765. His elopement in 1772 with Elizabeth Surtees—according to this letter she was also a pupil of Hutton—became a celebrated scandal, but Scott went on to a distinguished legal career, becoming Solicitor-General in 1788 and Lord Chancellor in 1801; in 1821 he became the first Earl of Eldon: ODNB; memoir of Eldon in *The Imperial magazine* 9 (July 1827), 592–9. For Henry Hutton see note to Letter 12.

[1] 1 February; Hutton died on 27 January.

[2] Hutton had written to Eldon to thank him for his subscription to the marble bust of himself, on which see note to Letter 125.

[3] Eldon's papers are voluminous and rather scattered, and I have not determined whether Hutton's letter is extant among them.

[4] Money having being left over after the bust was completed, a commemorative medal was struck to a design by Benjamin Wyon and sent to subscribers. It showed Hutton's profile on the obverse, and on the reverse stylized depictions of 'weighing the world' and of measuring the force of gunpowder: Royal Society image collection, IM/002330; Metropolitan Museum of Art, mma_charles_hutton_210249.

REFERENCES

Abell, Francis. *Prisoners of War in Britain 1756 to 1815* (Oxford, 1914).

Anonymous. *An Appeal to the Fellows of the Royal Society* (London, 1784).

Anonymous. *An History of the Instances of Exclusion from the Royal Society* (London, 1784).

Anonymous. *Supplement to the Appeal to the Fellows of the Royal Society* ([London], [1784]).

Anonymous. 'Doctor Charles Hutton', in: *Public Characters of 1799* (London, 1799), 97–123.

Anonymous. *A Catalogue of the Entire, Extensive and Very Rare Mathematical Library of Charles Hutton, L.L.D.* (London, [1816]).

Anonymous. *Tribute of Respect to Charles Hutton* ([London], [1822]).

Anonymous. *The Lunar Society: An Exhibition in Connection with the Bicentenary Celebrations of the Lunar Society of Birmingham Held in Collaboration with the Birmingham Reference Library, October 12th to November 26th, 1966* (Birmingham, 1966).

Archibald, Raymond Clare. 'Notes on some minor English mathematical serials', *The Mathematical Gazette* 14 (1929), 379–400.

Attinger, Gilles (ed.). *Dictionnaire historique de la Suisse* (Hauterive, 2004).

Austin, Gilbert. *Chironomia; or, A Treatise on Rhetorical Delivery* (London, 1806).

Baily, Francis. *The Doctrine of Interest and Annuities Analytically Investigated and Explained* (London, 1808).

Baily, Francis. 'On a method of fixing a transit instrument exactly in the meridian', *Memoirs of the Astronomical Society of London* 1 (1822), 59–66.

Bell, John. *Bell's Complete and Correct List of the Army* (London, 1782).

Benham, Daniel. *Memoirs of James Hutton* (London, 1856).

'Bey, Ali'. *Travels of Ali Bey in Morocco, Tripoli, Cyprus, Egypt, Arabia, Syria, and Turkey, between the Years 1803 and 1807* (London, 1816).

Bowdich, Thomas Edward. *Mission from Cape Coast Castle to Ashantee* (London, 1819).

Boyle, J.R. *Vestiges of Old Newcastle and Gateshead* (Newcastle, 1890).

Brewster, David. *Memoirs of the Life of Sir Isaac Newton* (London, 1875).

Brigden, Thomas E. 'Samuel Wesley, junior, and his circle, 1690–1739 [part III]', *Wesley Historical Society Proceedings* 11 (1918), 97–102.

Bright, Benjamin. *Catalogue of the ... Collection of Manuscripts Formed by ... Benjamin Heywood Bright ... Which Will Be Sold by Auction* (London, 1844).

Bright, Benjamin. *Catalogue of the Valuable Library of ... Benjamin Heywood Bright, Which Will Be Sold by Auction* (London, 1845).

Bruce, Gainsford. *The Life and Letters of John Collingwood Bruce of Newcastle-upon-Tyne* (Edinburgh and London, 1905).

Bruce, J. Collingwood. *A Hand-Book to Newcastle-on-Tyne* (London and Newcastle, 1863).

Bruce, John. *A Memoir of Charles Hutton* (Newcastle, 1823).

Cavendish, Henry. 'Experiments to determine the density of the Earth', *Philosophical Transactions* 88 (1798), 469–526.

Cazaud, Mr. 'Account of a new method of cultivating the sugar cane', *Philosophical Transactions* 69 (1779), 207–78.

Cecchi, Alberto. 'The "Arch of Equilibration" of Charles Hutton (1772)', *Meccanica* 45 (2010), 829–33.

Chambers, Ephraim. *Cyclopædia: Or, an Universal Dictionary of Arts and Sciences* (London, 1728).

Chambers, Neil. *Scientific Correspondence of Sir Joseph Banks, 1795–1820* (London, 2007).

Charles Grant, Vicomte de Vaux. *The Means of Finding the Longitude at Sea* (London, 1808).

Cook, James and King, James. *A Voyage to the Pacific Ocean* (London, 1784).

Costa, Shelley. 'The *Ladies' Diary*: society, gender and mathematics in England, 1704–1754', Ph.D. dissertation, Cornell University (2000).

Costa, Shelley. 'The *Ladies' Diary*: Gender, mathematics, and civil society in early-eighteenth-century England', *Osiris* 17 (2002), 49–73.

Craik, Alex D.D. 'Mathematical analysis and physical astronomy in Great Britain and Ireland, 1790–1831: some new light on the French connection', *Revue d'histoire des mathématiques* 22 (2016), 223–94.

Croarken, Mary. 'Providing longitude for all: The eighteenth century computers of the Nautical Almanac', *Journal of Maritime Research* 4 (2002), 106–26.

Croarken, Mary. 'Mary Edwards: Computing for a living in 18th-century England', *Annals of the History of Computing* 25 (2003), 9–15.

Croarken, Mary. 'Tabulating the Heavens: Computing the nautical almanac in 18th-century England', *Annals of the History of Computing* 25 (2003), 48–61.

Crosland, Maurice and Smith, Crosbie. 'The transmission of physics from France to Britain: 1800–1840', *Historical Studies in the Physical Sciences* 9 (1978), 1–61.

Dalby, Isaac. 'The longitudes of Dunkirk and Paris from Greenwich, deduced from the triangular measurement in 1787, 1788, supposing the Earth to be an ellipsoid', *Philosophical Transactions of the Royal Society of London* 81 (1791), 236–245.

Dauben, Joseph W. and Scriba, Christoph J. *Writing the History of Mathematics: Its Historical Development* (Basel, 2002).

Denniss, John. *Figuring It Out: Children's Arithmetical Manuscripts, 1680–1880* (Oxford, 2012).

Descartes, René. *La géométrie* (Leiden, 1637).

Despiau, M.L. *Select Amusements in Philosophy and Mathematics* (London, 1801).

Dickinson, Henry Winram. *Matthew Boulton* (Cambridge, 1936).

Duncan, Francis. *History of the Royal Regiment of Artillery* (London, 1879).

Eisenhart, Luther P. 'Walter Minto and the earl of Buchan', *Proceedings of the American Philosophical Society* 94 (1950), 282–94.

Ellerton, Nerida and Clements, M.A. *Rewriting the History of School Mathematics in North America 1607–1861: The Central Role of Cyphering Books* (Dordrecht, 2012).

Ellerton, Nerida and Clements, M.A. *Abraham Lincoln's Cyphering Book and Ten Other Extraordinary Cyphering Books* (Dordrecht, 2014).

Ellerton, Nerida and Clements, M.A. *Samuel Pepys, Isaac Newton, James Hodgson and the Beginnings of Secondary School Mathematics: A History of the Royal Mathematical School within Christ's Hospital, London 1673–1868* (Dordrecht, 2017).

Elsam, Richard. *An Essay on Rural Architecture* (London, 1803).

Emerson, William. *A Treatise of Algebra, in two books* (London, 1764, 1780).

Emerson, William. *Miscellanies. Or a Miscellaneous Treatise; Containing Several Mathematical Subjects* (London, 1776).

Erskine, David Steuart and Minto, Walter. *An Account of the Life, Writings, and Inventions of John Napier of Merchiston* (Perth, 1778, 1787).

Forbes, Eric G. 'Collections II: The Crawford collection of books and manuscripts on the history of astronomy, mathematics, etc., at the Royal Observatory, Edinburgh', *The British Journal for the History of Science* 6 (1973), 459–61.

Franklin, Benjamin and Franklin, William Temple. *The Private Correspondence of Benjamin Franklin* (London, 1817).

Franklin, Benjamin and Franklin, William Temple. *Memoirs of the Life and Writings of Benjamin Franklin* (London, 1818).

Fullom, Stephen Watson. *The Life of General Sir Howard Douglas* (London, 1863).

Galton, Francis. *Hereditary Genius* (London, 1870).

Gillispie, Charles Coulston, Holmes, Frederic Lawrence, and Koertge, Noretta. *Complete Dictionary of Scientific Biography* (Detroit, 2008).

Gingerich, Owen. 'William Herschel's 1784 autobiography', *Harvard Library Bulletin* 32 (1984), 73–82.

Goodricke, John. 'A series of observations on, and a discovery of, the period of the variation of the light of the bright star in the head of Medusa, called Algol', *Philosophical Transactions* 73 (1783), 474–82.

Goodricke, John. 'On the periods of the changes of light in the star Algol', *Philosophical Transactions* 74 (1784), 287–92.

[Gregory, Olinthus]. 'A review of some leading points in the official character and proceedings of the late President of the Royal Society', *London and Edinburgh Philosophical Magazine* 56 (1820), 161–74, 241–57.

Gregory, Olinthus. 'Brief memoir of the life and writings of Charles Hutton', *Imperial Magazine* 5 (1823), 201–27.

[Gregory, Olinthus]. 'Memoir of Charles Hutton, LL.D. F.R.S.', *The European Magazine* (June, 1823), 483–7.

Guicciardini, Niccolò. *The Development of Newtonian Calculus in Britain 1700–1800* (Cambridge, 1989).

Guicciardini, Niccolò. 'Dot-Age: Newton's mathematical legacy in the eighteenth century', *Early Science and Medicine* 9 (2004), 218–56.

Guillery, Peter and Saint, Andrew. *Woolwich (Survey of London, vol. 48)* (New Haven and London, 2012).

Hales, William. *Analysis Fluxionum* (London, 1800).

Halley, Edmund. 'An estimate of the degrees of the mortality of mankind', *Philosophical Transactions* 17 (1693), 596–610.

Halley, Edmund. 'Methodus Nova Accurata et Facilis Inveniendi Radices Æquationum Quarumcumque Generaliter, Sine Praevia Reductione', *Philosophical Transactions* 18 (1694), 136–48.

Hamilton, Hugh. *De sectionibus conicis. Tractatus geometricus. In quo, ex natura ipsius coni, sectionum affectioens [sic] facillime deducuntur* (London, 1758).

Hamilton, Hugh. *A Geometrical Treatise of the Conic Sections, in which the Properties of the Sections Are Derived from the Nature of the Cone, in an Easy Manner, and by a New Method* (London, 1773).

Hans, N. A. *New Trends in Education in the Eighteenth Century* (London, 1951).

Heilbron, John. 'A mathematicians' mutiny with morals', in: Horwich, Paul (ed.). *World Changes. Thomas Kuhn and the Nature of Science* (Cambridge, MA, 1993), 81–129.

Hellins, John. 'A new method of computing the value of a slowly converging series, of which all the terms are affirmative', *Philosophical Transactions* 88 (1798), 183–99.

Herschel, William. 'Astronomical observations relating to the mountains of the moon', *Philosophical Transactions* 70 (1780), 507–26.

Herschel, William. 'Astronomical observations on the rotation of the planets round their axes, made with a view to determine whether the Earth's diurnal motion is perfectly equable', *Philosophical Transactions* 71 (1781), 115–38.

Herschel, William. 'William Herschel', *European Magazine* 7 (January 1785), 1–3.

Higgitt, Rebekah (ed.). *Maskelyne: Astronomer Royal* (London, 2014).

Horsley, P.M. *Eighteenth-Century Newcastle* (Newcastle, 1971).

Horsley, Samuel. *Elementary Treatises on the Fundamental Principles of Practical Mathematics* (Oxford, 1801).

Horsley, Samuel. *Euclidis elementorum libri priores XII* (Oxford, 1802).

Horsley, Samuel ('Misogallus'). 'To Sir Joseph Banks, &c. &c. &c.', *Cobbett's Annual Register* (27 March 1802), 327–31.

Howse, Derek. *Nevil Maskelyne: The Seaman's Astronomer* (Cambridge, 1989).

Howson, A.G. *A History of Mathematics Education in England* (Cambridge, 1982).

Hues, Robert. *Tractatus de globis et eorum usu* (London, 1594).

Hues, Robert. *Tractatus duo mathematici. Quorum primus De globis coelesti et terrestri, eorum usu, a Roberto Hues Anglo, conscriptus. Alter Breviarium totius orbis terrarum, Petri Bertii, mathematici & professoris regii.* (Oxford, 1651).

Hutton, Catherine. *The Tour of Africa* (London, 1819).

Hutton, Catherine. *Reminiscences of a Gentlewoman of the Last Century* (Birmingham, 1891).

Hutton, Charles. *The School-Master's Guide: Or, a Complete System of Practical Arithmetic, Adapted to the Use of Schools* (Newcastle upon Tyne, 1764).

Hutton, Charles. *A Treatise on Mensuration, Both in Theory and Practice* (Newcastle and London, 1770).

Hutton, Charles. *A Plan of Newcastle Upon Tyne and Gateshead, Taken from an Accurate Survey, Finished in the Year 1770* (Newcastle, 1772).

Hutton, Charles. *The Principles of Bridges* (Newcastle, London, and Edinburgh, 1772, 1801).

Hutton, Charles. *The Diarian Miscellany* (London, 1775).

Hutton, Charles. 'A demonstration of two theorems mentioned in Art. XXV. of the Philosophical Transactions for the Year 1775', *Philosophical Transactions* 66 (1776), 600–3.

Hutton, Charles. 'A new and general method of finding simple and quickly-converging series; by which the proportion of the diameter of a circle to its circumference may easily be computed to a great number of places of figures', *Philosophical Transactions* 66 (1776), 476–92.

Hutton, Charles. 'An account of the calculations made from the survey and measures taken at Schehallien, in order to ascertain the mean density of the Earth', *Philosophical Transactions* 68 (1778), 689–788.

Hutton, Charles. 'The force of fired gun-powder, and the initial velocities of cannon balls, determined by experiments', *Philosophical Transactions* 68 (1778), 50–85.

Hutton, Charles. 'Calculations to determine at what point in the side of a hill its attraction will be the greatest', *Philosophical Transactions* 70 (1779), 1–14.

Hutton, Charles. 'Of cubic equations and infinite series', *Philosophical Transactions* 70 (1780), 387–450.

Hutton, Charles. *Tables of the Products and Powers of Numbers* (London, 1781).

Hutton, Charles. 'Project for a new division of the quadrant', *Philosophical Transactions* 74 (1784), 21–34.

Hutton, Charles. *Mathematical Tables: Containing Common, Hyperbolic, and Logistic Logarithms* (London, 1785).

Hutton, Charles. *Elements of Conic Sections; with Select Exercises in Various Branches of Mathematics and Philosophy. For the Use of the Royal Military Academy at Woolwich* (London, 1787).

Hutton, Charles. 'Abstract of experiments made to determine the true resistance of the air to the surfaces of bodies, of various figures, and moved through it with different degrees of velocity', *Transactions of the Royal Society of Edinburgh* 2 (1790), 29–36.

Hutton, Charles. 'Experiments on the expansive force of freezing water, made by Major Williams at Quebec, in the years 1784 and 1785. Communicated in a letter from Charles Hutton', *Transactions of the Royal Society of Edinburgh* 2 (1790), 23–8.

Hutton, Charles (ed.). *The Works of John Whitehurst*, (London, 1792).

Hutton, Charles. *A Mathematical and Philosophical Dictionary* (London, 1795, 1796).

Hutton, Charles. *A Course of Mathematics* (London, 1798).

Hutton, Charles. 'To the Editor of the Monthly Review', *Monthly Review* 38 (June 1802), 222–4.

Hutton, Charles. 'To the Editor of the Monthly Magazine', *Monthly Magazine* (August 1802), 27–31.

Hutton, Charles. *Recreations in Mathematics and Natural Philosophy* (London, 1803, 1814).

Hutton, Charles, Shaw, George and Pearson, Richard. *The Philosophical Transactions of the Royal Society of London, from their Commencement, in 1665, to the Year 1800; Abridged, with Notes and Biographic Illustrations* (London, 1809).

Hutton, Charles. 'Note on the divining rod', *Philosophical Magazine* 55 (June 1820), 465–8.

Hutton, Charles. 'On the mean density of the Earth', *Philosophical Transactions* 111 (1821), 276–92.

Hutton, Charles. 'To the Editor of the Newcastle Magazine', *The Newcastle Magazine* (March 1822), 127–8.

Jack, Richard. *Elements of Conic Sections in Three Books: in Which Are Demonstrated the Principal Properties of the Parabola, Ellipse, and Hyperbola* (Edinburgh, 1742).

Jewitt, Llewellynn. 'Pedigree of the family of Hutton, of Derby, Birmingham', *The Reliquary* 11 (1871), foldout before p. 215.

Johnson, W. 'Charles Hutton, 1737–1823: The prototypical Woolwich Professor of mathematics', *Journal of Mechanical Working Technology* 18 (1989), 195–230.

Johnson, W. *Collected Works on Benjamin Robins and Charles Hutton* (New Delhi, 2001).

Johnson, W.M and Exley, Thomas. 'Hutton, Charlotte', in: *Imperial Encyclopaedia* (London, 1812), vol. 3, pp. 60–1.

Jones, William. *Synopsis Palmariorum Matheseos: or, A New Introduction to the Mathematics* (London, 1706).

Jones, William Daniel. *Records of the Royal Military Academy* (Woolwich, 1851).

Jungnickel, Christa and McCormmach, Russell. *Cavendish: The Experimental Life*, 2nd edition (Lewisburg, PA, 1998).

Kane, John. *List of Officers of the Royal Regiment of Artillery* (Greenwich, 1815).

Kippis, Andrew. *Observations on the Late Contests in the Royal Society* (London, 1784).

Lalande, J. J. 'Notice sur la vie de Condorcet', *Mercure de France* (20 January 1796), 143.

Landen, John. *Mathematical Memoirs Respecting a Variety of Subjects; with an Appendix Containing Tables of Theorems for the Calculation of Fluents* (London, 1780).

Laplace, Pierre Simon de. *Traité de mécanique céleste* (Paris, 1799–1825).

Laplace, Pierre Simon de. 'Sur la figure de la Terre, et la loi de la pesanteur à sa surface', *Connaissance des Tems* (1821), 326–31.

Lemprière, William. *A Tour from Gibraltar to Tangier, Sallee, Mogodore, Santa Cruz, Tarudant; and Thence, over Mount Atlas, to Morocco* (London, 1791).

Levere, T.H and Turner, G.L'E. *Discussing Chemistry and Steam: The Minutes of a Coffeehouse Philosophical Society, 1780–1787* (Oxford, 2002).

Lexel, Mr. 'Extract of a letter from Mr. Lexel to Dr. Morton. Dated Petersburg, June 14, 1774', *Philosophical Transactions* 65 (1775), 280–2.

Leybourn, Thomas. *The Mathematical Questions, Proposed in the Ladies' Diary, and their Original Answers, Together with Some New Solutions, from 1704 to 1816* (London, 1817).

Leybourn, Thomas. 'Charles Hutton, LL.D. F.R.S.'. *The Mathematical Repository* new series, vol. 5 (1830), 187–96.

l'Hôpital, Guillaume de. *An Analytick Treatise of Conick Sections, and their Use for Resolving of Equations in Determinate and Indeterminate Problems* (London, 1723).

Lorgna, Antonio Mario. *A Dissertation on the Summation of Infinite Converging Series with Algebraic Divisors*, edited by Clarke, H. (London, 1779).

Mackenzie, Eneas. *Historical Account of Newcastle-Upon-Tyne Including the Borough of Gateshead* (Newcastle, 1827).

MacLaurin, Colin. *A Treatise of Fluxions* (Edinburgh, 1742).

Martin, Benjamin. *Miscellaneous Correspondence* (London, 1764).

Maseres, Francis. *A Dissertation on the Use of the Negative Sign in Algebra* (London, 1758).

Maseres, Francis. *Scriptores Logarithmici; or, a Collection of Several Curious Tracts on the Nature and Construction of Logarithms, Mentioned in Dr. Hutton's Historical Introduction to his New Edition of Sherwin's Mathematical Tables: Together with Some Tracts on the Binomial Theorem and Other Subjects Connected with the Doctrine of Logarithms* (London, 1791–1807).

Maseres, Francis and Babbage, Charles. *Scriptores Optici; or, a Collection of Tracts Relating to Optics* (London, 1823).

Maskelyne, Nevil. 'An account of observations made on the Mountain Schehallien for finding its attraction.', *Philosophical Transactions* 65 (1775), 500–42.

Maskelyne, Nevil. *Astronomical Observations Made at the Royal Observatory at Greenwich* (London, 1776–1811).

Mason, Charles. *Mayer's Lunar Tables, Improved by Mr. Charles Mason* (London, 1787).

'Mathers, Patrick'. *The Great and New Art of Weighing vanity, or, A Discovery of the Ignorance and Arrogance of the Great and New Artist, in his Pseudo-philosophical Writings by M. Patrick Mathers, Arch-Bedal to the University of S. Andrews ; to Which Are Annexed Some Tentamina de motu penduli & projectorum.* (Glasgow, 1672).

Maty, Paul Henry. *An Authentic Narrative of the Dissensions and Debates in the Royal Society* (London, 1784).

May, Leonard Morgan. *Charlton: Near Woolwich, Kent. Full and Complete Copies of All the Inscriptions in the Old Parish Church and Churchyard, Together with Notes on the History of the Manor and of the Families Connected with the Place* (London, 1908).

Melmore, Sidney. 'Some letters from Charles Hutton to Robert Harrison', *The Mathematical Gazette* 30 (1946), 71–81.

Montucla, Jean Étienne. *Histoire des mathématiques* (Paris, 1799–1802).

Morrell, J. B. 'The Leslie affair: careers, kirk and politics in Edinburgh in 1805', *The Scottish Historical Review* 54 (1975), 63–82.

Museum of Artillery. *Catalogue of the Museum of Artillery in the Rotunda at Woolwich* ([Manchester], 1963).

Myers, Robin. 'The Records of the Stationers' Company 1554–1920', (115 microfilms) (1986).

Napier, John. *Mirifici logarithmorum canonis descriptio* (Edinburgh, 1614).

Napier, John and Briggs, Henry. *Mirifici logarithmorum canonis constructio* (Edinburgh, 1619).

Newton, Isaac. *Philosophiæ naturalis principia mathematica* (London, 1687).

Newton, Isaac. *Isaaci Newtoni opera quæ exstant omnia*, edited by Horsley, Samuel (London, 1779–1785).

Matthew, H.C.G., Harrison, Brian, and Goldman, Lawrence. *The Oxford Dictionary of National Biography* (Oxford, 2004).

Pedersen, Olaf. 'The "Philomath" of 18th century England', *Centaurus* 8 (1963), 238–62.

Perl, Teri. 'The Ladies' Diary or Woman's Almanack, 1704–1841', *Historia Mathematica* 6 (1979), 36–53.

Playfair, John. 'On the causes which affect the accuracy of barometrical measurements', *Transactions of the Royal Society of Edinburgh* 1 (1788), 87–130.

Playfair, John. 'An account of the lithological survey of Schehallion', *Philosophical Transactions* 101 (1811), 347–78.

Playfair, J.G. (ed.). *The Works of John Playfair* (Edinburgh, 1822).

Powell, William Samuel. *A Defence of the Observations on the First Chapter of a Book Called Miscellanea Analytica* (London, 1760).

Powell, William Samuel. *Observations on the First Chapter of a Book Called Miscellanea Analytica* (Cambridge, 1760).

Pringle, John. *A Discourse on the Theory of Gunnery. Delivered at the Anniversary Meeting of the Royal Society, November 30, 1778* (London, 1778).

Rackwitz, Joseph Friedrich Freiherr von. *Ueber den Schachspieler des herrn von Kempelen* (Leipzig and Dresden, 1789).

Raphson, John. *Analysis aequationum universalis* (London, 1690).

Robertson, Abram. *Sectionum conicarum libri septem* (Oxford, 1792).

Robins, Benjamin. *New Principles of Gunnery ... A New Edition*, edited by Hutton, Charles (London, 1805).

Robinson, F.J.G. 'Trends in education in Northern England in the eighteenth century', Ph.D. dissertation, University of Newcastle-upon-Tyne (1972).

Rogers, Charles. 'Register of the Collegiate Church of Crail, Fifeshire. With historical remarks', *Transactions of the Royal Historical Society* 6 (1877), 324–94.

Saint, William. *Four Letters to Lieutenant Colonel Mudge* (London, 1811).

Scott, Walter. *The Pirate* (Edinburgh, 1822).

Senex. 'Board of ordnance in other days', *Mechanics' Magazine* 55 (1851), 324–9.

Seymour, W.A and Andrews, J.H. *A History of the Ordnance Survey* (Folkestone, 1980).

Simpson, Thomas. *The Doctrine of Annuities and Reversions, Deduced from General and Evident Principles* (London, 1742).

Simpson, Thomas. *Mathematical Dissertations on a Variety of Physical and Analytical Subjects* (London, 1743).

Simpson, Thomas. *Elements of Plane Geometry* (London, 1747).

Simpson, Thomas. *The Doctrine and Application of Fluxions* (London, 1750).

Simpson, Thomas. *Select Exercises for Young Proficients in the Mathematicks* (London, 1752).

Simpson, Thomas. *Select Exercises for Young Proficients in the Mathematicks ... A New Edition*, edited by Hutton, Charles (London, 1792).

Simson, Robert. *Sectionum conicarum libri V* (Edinburgh, 1735).

Simson, Robert. *Elements of the Conic Sections* (Edinburgh, 1775).

Skempton, A.W and Chrimes, Mike. *Biographical Dictionary of Civil Engineers in Great Britain and Ireland. Vol. 1, 1500–1830* (London, 2002).

Smiles, Samuel. *Lives of Boulton and Watt* (London, 1865).

Smith, George. *An Universal Military Dictionary* (London, 1779).

Smith, Robert. *A Compleat System of Opticks* (Cambridge, 1738).

Smith, Robert. *Harmonics, or the Philosophy of Musical Sounds* (Cambridge, 1749).

'Snip, Simon'. *The Philosophical Puppet Show* ([London], [1783?]).

Southgate and Son. *Catalogue of a Miscellaneous Collection of Books: Being the Valuable and Scientific Library of the Late Dr. Olinthus Gregory ...: Which Will Be Sold by Auction by Messrs. Southgate and Son ... on Thursday, March the 17th, 1842 and Following Day* ([London], [1842]).

Stirling, James. *Methodus differentialis: sive tractatus de summatione et interpolatione serierum infinitarum* (London, 1730).

[Strange, John]. *Canons of Criticism, Extracted from the Beauties of Maty's Review* (London, 1784).

Swetz, Frank J. 'The mystery of Robert Adrain', *Mathematics Magazine* 81 (2008), 332–44.

Taylor, E.G.R. *The Mathematical Practitioners of Hanoverian England, 1714–1840* (Cambridge, 1966).

Taylor, Michael. *Table of Logarithms ... of the Sines and Tangents to Every Second of the Quadrant ... With a Preface and Precepts for the Explanation and Use of the Same, by Nevil Maskelyne, F.R.S. Astronomer Royal* (London, 1792).

'The Vignoles Snuff-Box', *Notes and Records of the Royal Society of London* 4 (1946), 45–48.

Vallemont, Pierre Le Lorrain de. *La physique occulte, ou, Traité de la baguette divinatoire* (Paris, 1693).

Vibart, H. M. *Addiscombe, its Heroes and Men of Note* (Westminster, 1894).

Vignoles, Keith H. *The Infant Ensign: The Story of Charles and Camilla Vignoles and their Son Charles Blacker* (Emsworth, 1967).

Vignoles, Keith Hutton. *Charles Blacker Vignoles, Romantic Engineer* (Cambridge, 1982).

Vignoles, Olinthus J. *Life of Charles Blacker Vignoles: A Reminiscence of Early Railway History* (London, 1889).

Vince, Samuel. *The Elements of the Conic Sections, as Preparatory to the Reading of Sir I. Newton's Principia* (Cambridge, 1781).

Walkingame, Francis. *The Tutor's Assistant* (London, 1751).

Wallis, John. *A Treatise of Algebra, Both Historical and Practical* (London, 1685).

Wallis, Peter John. 'Newcastle Mathematical Libraries: William Armstrong, Charles Hutton and others', *Northern Notes* 4 (1972) (supplement).

Wallis, Peter John. 'The book subscription lists project: its relevance for historians of mathematics', *Historia Mathematica* 2 (1975), 321–6.

Wallis, Peter John and Wallis, Ruth. *Mathematical Tradition in the North of England* (Durham, 1991).

Wallis, Peter John. *Index of British Mathematicians. Part III, 1701–1800* (Newcastle upon Tyne, 1993).

War Office. *A List of All the Officers of the Army and Royal Marines on Full and Half-Pay* (1814).

Wardhaugh, Benjamin. 'Charles Hutton and the "Dissensions" of 1783–84: scientific networking and its failures', *Notes and Records of the Royal Society* 71 (2017), 41–59.

Wardhaugh, Benjamin. 'Charles Hutton, "One of the greatest mathematicians in Europe"?', *BSHM Bulletin* 32 (2016), 91–99.

Wardhaugh, Benjamin. *Gunpowder and Geometry: The Remarkable Life of Charles Hutton (1737–1823)* (forthcoming, 2018).

Waring, Edward. *A Letter to the Reverend Dr. Powell* (Cambridge, 1760).

Waring, Edward. *On the Principles of Translating Algebraic Quantities into Probable Relations and Annuities* (Cambridge, 1792).

Waring, Edward. 'Original Letter of the Late Dr. Waring', *Monthly Magazine* 7 (1799), 306–10.

Watson, Robert Spence. *The History of the Literary and Philosophical Society of Newcastle-upon-Tyne* (London, 1897).

Watts, I.P. '"We want no authors": William Nicholson and the Contested Role of the Scientific Journal in Britain, 1797–1813', *British Journal for the History of Science* 47 (2014), 397–419.

Weatherford, John. 'The Vicomte De Vaux: would-be Canadian', *Ontario History* 47 (1955), 49–51.

Weld, Charles Richard. *Descriptive Catalogue of the Portraits in the Possession of the Royal Society* (London, 1860).

Welford, R. *Men of Mark 'twixt Tyne and Tweed* (London, 1895).

Whitehurst, John. *An Attempt towards Obtaining Invariable Measures of Length* (London, 1787).

Wigham, H.M. 'A bag of old letters: Thomas Collinson', *Friends' Quarterly Examiner* 25 (1891), 406–16.

Wilkinson, T.T. 'Mathematics and Mathematicians, the Journals of the Late Reuben Burrow', *London, Edinburgh, and Dublin Philosophical Magazine* 4th series, vol. 5 (1853), 185–93, 514–22, and (vol. 6) 196–204.

Wilson, John. *A Vindication of the Miscellanea Analytica* (Cambridge, 1760).

Winterburn, Emily. 'Philomaths, Herschel, and the Myth of the Self-taught Man', *Notes and Records of the Royal Society of London* 68 (2014), 207–25.

Wolcot, John ('Peter Pindar'). *Peter's Prophecy* (Dublin, 1789).

Wood, James. *The Elements of Algebra: Designed for the Use of Students in the University* (Cambridge, 1795).

INDEX

The manufacturer's authorised representative in the EU for product safety is Oxford University Press España S.A. of el Parque Empresarial San Fernando de Henares, Avenida de Castilla, 2 – 28830 Madrid (www.oup.es/en or product.safety@oup.com). OUP España S.A. also acts as importer into Spain of products made by the manufacturer.

www.ingramcontent.com/pod-product-compliance
Lightning Source LLC
LaVergne TN
LVHW011755200926
848484LV00003B/709
* 9 7 8 0 1 9 8 8 0 5 0 4 5 *